Pocket Guide to Instrumentation

Pocket Guide to Instrumentation

R. R. Lee

Gulf Publishing Company
Houston, Texas

To my wife,
Patricia M. Lee

Pocket Guide to Instrumentation

Gulf Publishing Company
Book Division
P.O. Box 2608 □ Houston, Texas 77252-2608

10 9 8 7 6 5 4 3 2 1

Library of Congress Cataloging-in-Publication Data

Lee, R. R. (Robert R.), 1928—
Pocket guide to instrumentation / R. R. Lee.
p. cm.
ISBN 0-88415-308-8 (alk. paper)
1. Automatic control Handbooks, manuals, etc. 2. Engineering instruments Handbooks, manuals, etc. I. Title.
TJ225.L43 1999
629.8—dc21 99-41792
CIP

Printed in the United States of America.

Printed on acid-free paper (∞).

Contents

Acknowledgments

I would like to express my appreciation to certain people, manufacturers and former co-workers, for support and contributions in the preparation of this book: L. L. Lee, G. Hale, Gulf Coast Oil & Gas Industries; J. M. Smith, J. E. Hardwick, C. L. Davis, MIS-U.A.E.; Roger Young, Das Nair, Mohd Razzaq, ARCO Dubai Inc.; Jill Maple, Duracell Inc.; Ed Shreeve, Dick van Gallen Last, Ras Al Khaimah Gas Commission-U.A.E.; W. D. Wilson, Crawford Fitting Co.; H. H. Hittner, Yarway Corp.; K. D. Arndt, Jr., Keystone Valve Co.; Jeffrey A. Merchant, Jaqua-McKee; R. T. Burnstad Jr., The Walworth Company.

I especially want to remember Frank Weirch for the ideas he contributed to this book.

Preface

This pocket guide is designed and intended to assist materials personnel in handling project instrumentation equipment and instrument fittings, valves, cable, and many other key components used in industrial plants. The many pictures, charts, tables, and other data serve as convenient references for identifying and describing the materials that are typical of such projects, and which are often the responsibility of various disciplines.

It is practically impossible to cover all instrument items for the varied applications, but this guide provides an overview of most of the components. It describes in detail instrument specialty valves along with the major valves used in process systems, such as gate, globe, check, ball, plug, and butterfly valves, in addition to some of the mini-types. In addition, this guide covers various styles of steam traps, tracings, panels, and panel instruments. It illustrates instrument fittings in a cross-referenced chart that covers five major fittings vendors. Gauges, thermowelds, thermocouples, thermometers, and other unique product detectors are also included.

This guide should be quite useful to drafters, purchasing agents, clerks, technicians, students, professors and project managers. The material in the book is believed to be technically correct; neither the author nor Gulf Publishing Company warrants its use. Always consult the applicable vendor's catalog or specifications issued by your project manager.

R. R. Lee
Houston, Texas

1
Control Devices

A materials person receiving and issuing project instrumentation materials is deluged with an array of new and varied instrument items, but at the same time, he or she continues to receive items for older process sections of plants that may operate on air or gas. So, careful checking of all parts and instruments is still the prime requirement.

TEMPERATURE AND PRESSURE TRANSMITTERS

Today's inventory can include unusual instruments such as temperature recorders for low-temperature service inside buildings, refrigerators or ovens; and spring-wound clock-driven circular chart recorders that never require ink or any type of power, but simply are rewound when the chart is changed (Figure 1-1). And there are many variations of temperature transmitters and recorders, such as temperature recorders capable of programmable inputs, speeds, and alarms, and some have two alarm settings per pen, both high and low with digital displays. These circular recorders are capable of recording up to 2,300°C (Figure 1-2).

It is not the job of the materials person to know how these instruments operate, or which one to install. The materials person's job is to control and protect these units from damage during storage at the project warehouse until ordered installed by the engineers or the project manager.

Figure 1-1. Chart-Temp® recorder, key wound. (*©Copyright Omega Engineering, Inc. All rights reserved. Reproduced with permission of Omega Engineering, Inc., Stamford, CT 06907.*)

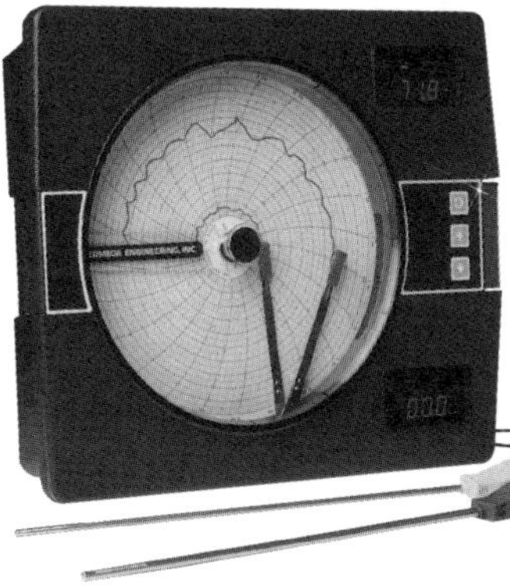

Figure 1-2. Circular 2-pen recorder. (*©Copyright Omega Engineering, Inc. All rights reserved. Reproduced with permission of Omega Engineering, Inc., Stamford, CT 06907.*)

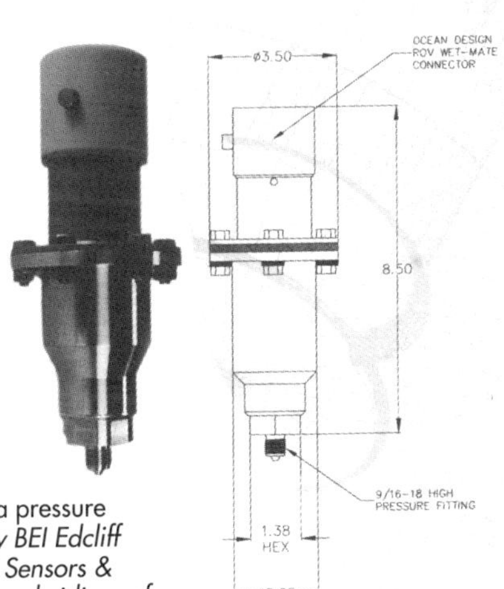

Figure 1-3. Subsea pressure transmitter. (*Courtesy BEI Edcliff Instruments Division, Sensors & Systems Company, a subsidiary of BEI Technologies, Inc.*)

Figure 1-3 shows an unusual pressure transmitter that operates in 10,000 ft of saltwater, withstands 35,000 psi internal failure pressure, and survives and operates on the seafloor for 20 years.

Figure 1-4 is another pressure transmitter built for rugged service and transmits flow levels or pressures of 0–1,000 psid.

New instrumentation products continue to appear at job sites. Colored tubing and hose, including conduit, are seen

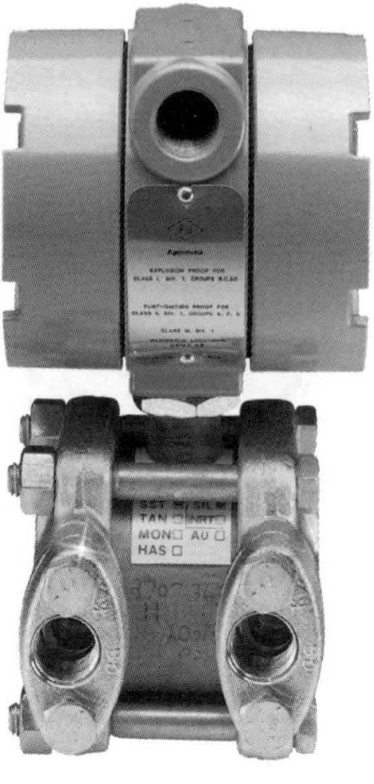

Figure 1-4. Heavy-duty process transmitter. (*©Copyright Omega Engineering, Inc. All rights reserved. Reproduced with permission of Omega Engineering, Inc., Stamford, CT 06907.*)

often at job sites, as are static mixer assemblies for wastewater applications and chemical mixing inside a hose or pipe (Figure 1-5). Large display panel meters with 4-in.-high digits are used. (Figure 1-6). Just as important, and still used in many plants are earlier modes of panel board and digital indicators (Figure 1-7).

Intrinsically safe field temperature transmitters (Figure 1-8) are used in Class 1, Division 1, Groups A, B, C, D. Intrinsically safe is defined as "equipment and wiring that are incapable of releasing sufficient electrical energy to cause ignition of a specific hazardous mixture in its most easily ignited concentration." In plain terms, to have a fire or explosion, a source of ignition must be present. An intrinsically safe system assumes the source is present in the atmosphere, but the system is designed so that electrical or thermal energy of a particular instrument loop can never be great enough to cause ignition. See Figure 1-9 for the causes of combustion.

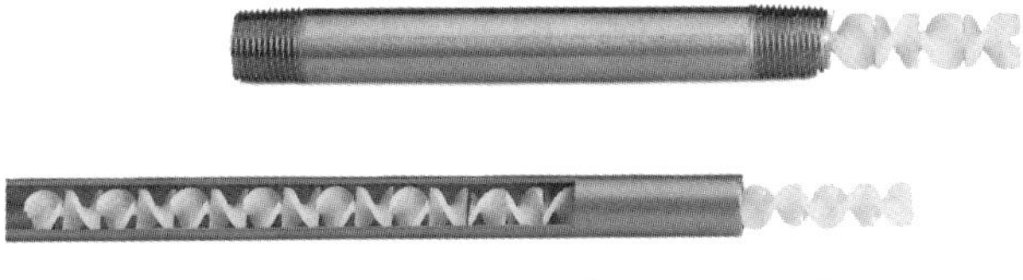

Figure 1-5. 304SS static mixer assemblies. (*©Copyright Omega Engineering, Inc. All rights reserved. Reproduced with permission of Omega Engineering, Inc., Stamford, CT 06907.*)

Figure 1-6. Large display (2–4 in.) digital panel meters. (*©Copyright Omega Engineering, Inc. All rights reserved. Reproduced with permission of Omega Engineering, Inc., Stamford, CT 06907.*)

INSTRUMENT INDICATING DEVICES

Figure 1-7 illustrates control panel indicating devices by Foxboro. Figure 1-10 shows an electronic temperature indicating transmitter. The probe has a ¾-in. bushing that will screw directly into a thermowell outlet on a vessel or into a process pipe. There are many variations of temperature indicators and transmitters. This particular transmitter has a range of 450 to 1,000°F.

Figure 1-11A illustrates a liquid level controller/transmitter that is top mounted. A DC power supply is required on the Fisher electronic transmitter. It attaches to an opening in a vessel by a flange, gasket, and bolts. The typical level controller/transmitter shown in Figure 1-11B is an external displacement type with top and bottom pipe connections.

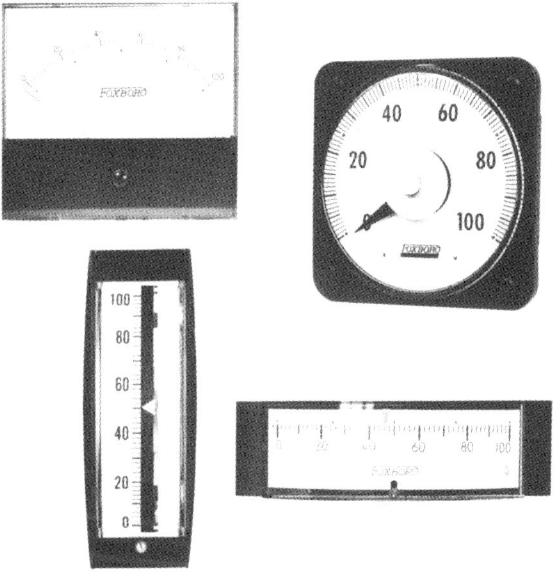

Figure 1-7. Control panel indicating devices. (*Courtesy of Richardson Engineering Services, Inc.*)

Metering toxic fluids such as nitric and sulfuric acids, boiling caustics, wet chlorine at high temperatures and high pressures is performed by straight-through rotameters. The Wallace & Tiernan Varea-meters™ operate by a float magnet and a follower magnet on the pointer. At 50% on the scale (see Figure 1-12), unlike poles attract equally so that

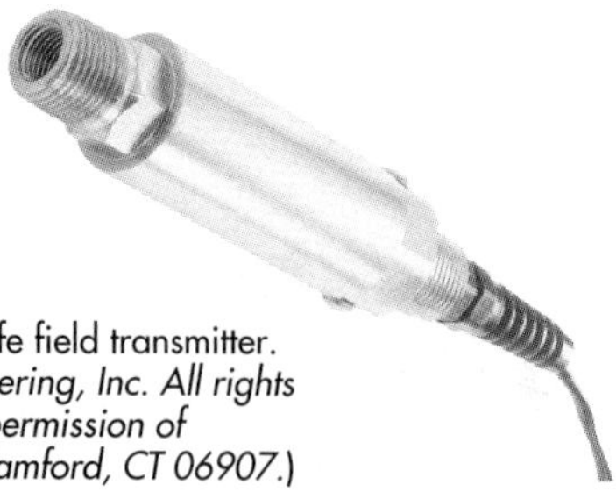

Figure 1-8. Intrinsically safe field transmitter. (*©Copyright Omega Engineering, Inc. All rights reserved. Reproduced with permission of Omega Engineering, Inc., Stamford, CT 06907.*)

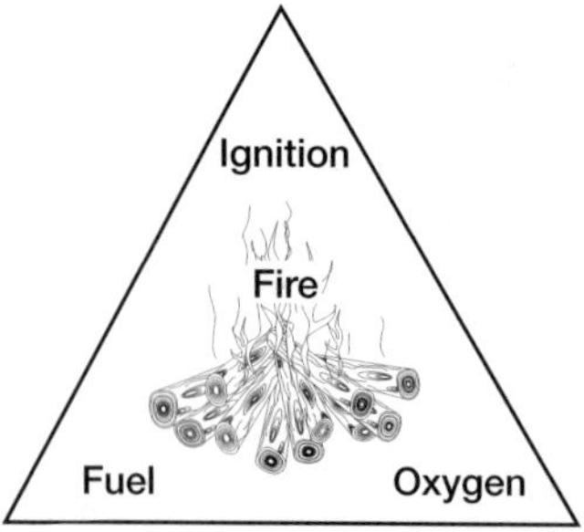

Figure 1-9. Causes for combustion. (*©Copyright Omega Engineering, Inc. All rights reserved. Reproduced with permission of Omega Engineering, Inc., Stamford, CT 06907.*

the rotating magnet is parallel to the float magnet. Figure 1-13 shows a Varea-meter™ that is manufactured in carbon steel, stainless steel, and PVC. For service in hazardous locations, an explosion-dust-ignition-proof thick-walled

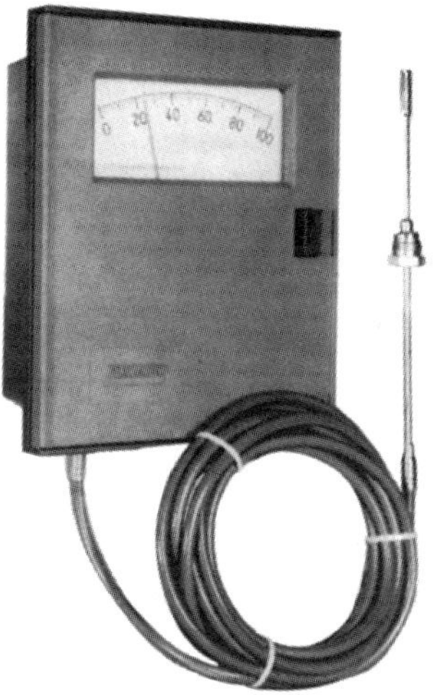

Figure 1-10. Temperature indicating transmitter. *(Courtesy of Richardson Engineering Services, Inc.)*

aluminum enclosure with a heavy-duty glass is available, as shown in Figure 1-14.

In specific hazardous atmospheres in oil refineries, chemical plants, paint manufacturing plants, metal finishing areas, coal processing locations, and grainaries, thermostats are used to control the line voltage equipment for heating, air-conditioning, and refrigeration. Figure 1-15 illustrates a thermostat with Mercoid control. For refrigeration and air-conditioning, the switch contacts close on temperature rise, whereas for heating, they close on temperature fall. A removable knob is used to set the temperature setting. The ambient room temperature is indicated on the front.

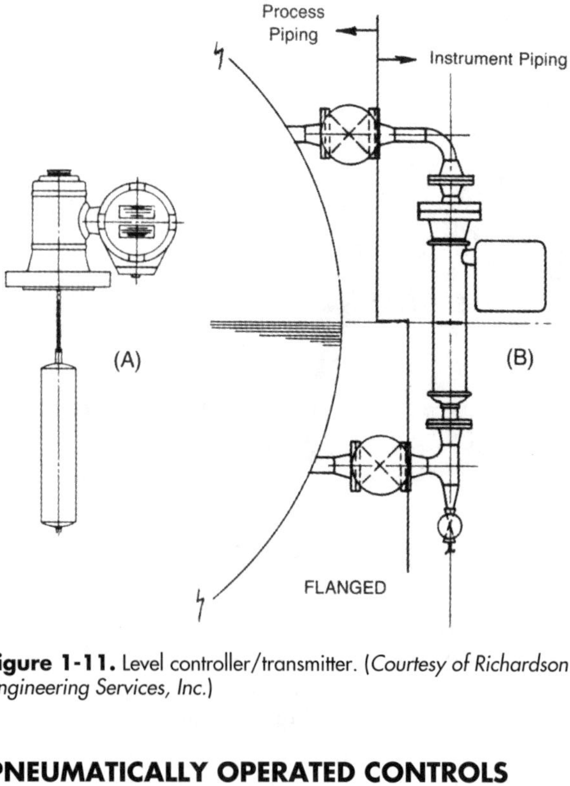

Figure 1-11. Level controller/transmitter. (*Courtesy of Richardson Engineering Services, Inc.*)

PNEUMATICALLY OPERATED CONTROLS

Many plants use and will continue to use pneumatic controls, therefore it is important to be familiar with pneumatic materials. Figure 1-16 shows a liquid level transmitter (A),

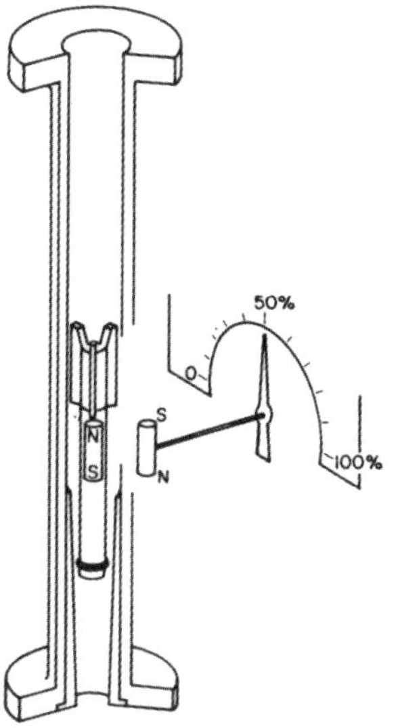

Figure 1-12. Magnetically operated rotameter. (*Courtesy of Wallace & Tiernan Div., Pennwalt Corp.*)

a pressure regulator (B), a backflow preventer (C), and a temperature regulator (D).

Figure 1-17 shows a pneumatic temperature transmitter (A), a pneumatic indicating pressure transmitter (B), and pneumatic absolute pressure transmitter (C), and a pneumatic flow transmitter (D).

Figure 1-18 illustrates a pneumatic flow indicator that totals actual flow on a six-digit counter from any flow transmitter (A), an air switch (B), and an indicator receiver gauge (C).

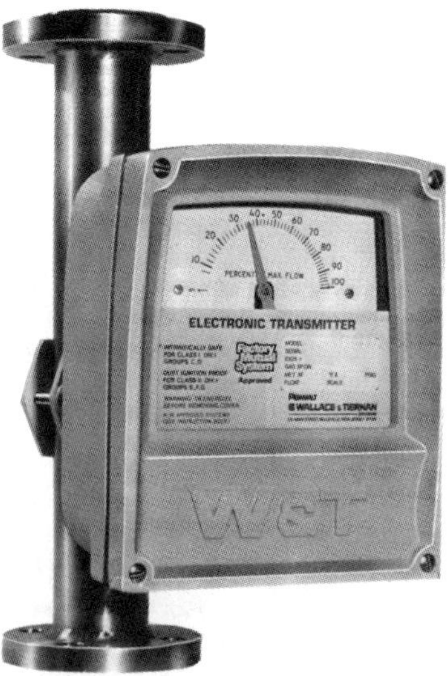

Figure 1-13. Varea-meter™ for non-hazardous areas. (*Courtesy of Wallace & Tiernan Div., Pennwalt Corp.*)

INSTRUMENT TUBING

Air or some form of gas is used to activate pneumatic instruments, and instrument tubing transports it. The tubing is manufactured in copper, PVC, stainless steel, etc. Figure 1-19 shows three different types of single-line tubing used in pneumatic and low-pressure hydraulic installations where flexible connections are needed. They are:

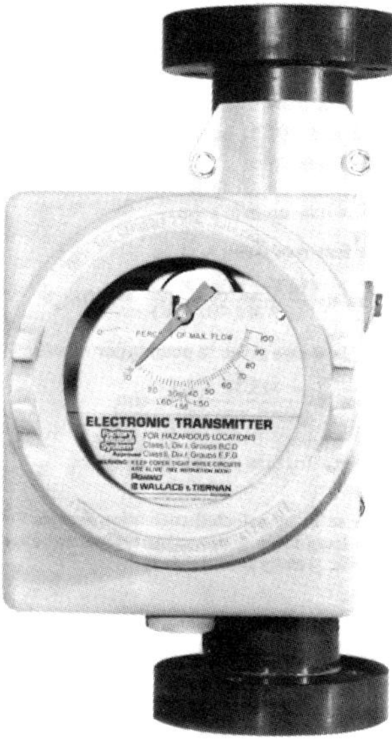

Figure 1-14. Varea-meter™ for hazardous areas. (*Courtesy of Wallace & Tiernan Div., Pennwalt Corp.*)

1. "P" tubing (polyethylene) tubing, which is used in low-pressure installations.
2. Dekoron® "N" tubing, which is used where maximum mechanical strength is required for high temperature and pressure.
3. Dekoron® FR (flame retardant) tubing, which is used where a plastic tubing with a flame retardant characteristic is desired.

Figure 1-15. Thermostat for hazardous areas. (*Courtesy of Crouse-Hinds Co.*)

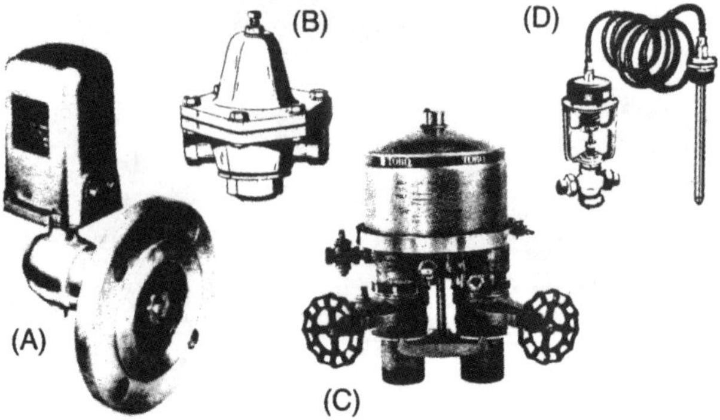

Figure 1-16. Pneumatic controls. (*Courtesy of Richardson Engineering Services, Inc.*)

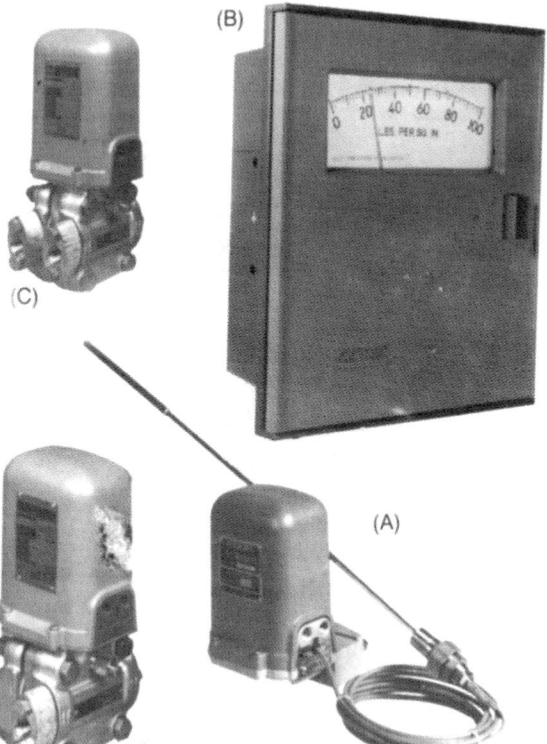

Figure 1-17. Pneumatic controls. (*Courtesy of Richardson Engineering Services, Inc.*)

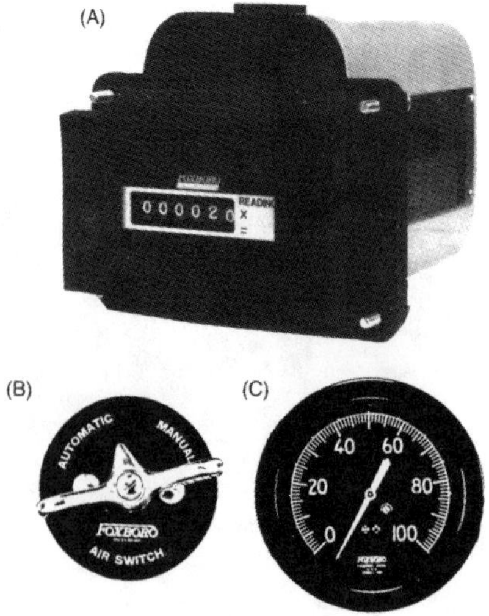

Figure 1-18. Pneumatic panel-mounted controls. (*Courtesy of Richardson Engineering Services, Inc.*)

Packing of single-line tubing is shown in Figure 1-20. The numerals 1–30 are printed the full length of the tubing. These markings identify each end at a given point. By cutting the lead end at a given number, and marking the reel or carton by the same number, tubes can be easily identified for the proper connection or for materials control purposes in the project warehouse.

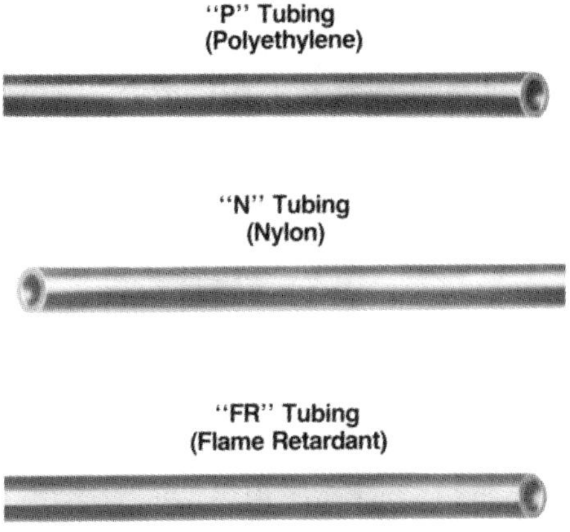

Figure 1-19. Flexible instrument tubing. (*Courtesy of Eaton Corp., Industrial Polymer Products Div.*)

ARMORED METAL TUBE BUNDLES

Copper or aluminum tubing is packaged, in some cases, in an armored tubing bundle, suitable for direct burial, designed for tight bends, resistant to chemical attacks, and flame retardant. The individual tubes are number coded, and a 22-gauge communication conductor is included to help coordinate bundle installation. There are many variations of multiple tube bundles. Figure 1-21 shows Dekoron® Type

Shipping Weights and Lengths

Dekoron "P", "HP", "FR"			
TUBE SIZE (O.D.)	COIL LENGTH (Feet)	MASTER PACK (Feet)	ESTIMATED MASTER PACK SHIPPING WEIGHT (lbs.)
5/32"	500	2000	12.5
1/4"	250	1000	16.8
	500	1000	15.3
	1000	2000	25.1
3/8"	250	1000	32.2
	500	1000	30.2
1/2"	250	500	22.6
DEKORON "N"			
1/4"	250	1000	19.0
3/8"	250	1000	41.0

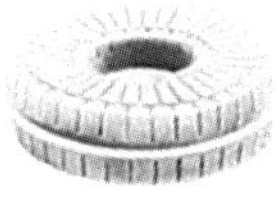

YO-YO PAK
This Dekoron exclusive keeps 250′ of 1/4" "P" or "FR" tubing, or 500′ of 5/32" "FR" tubing clean, and dry.

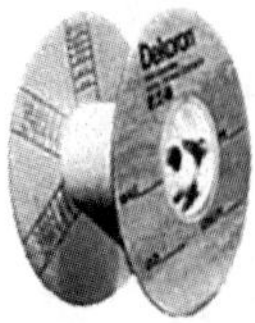

CONVENIENT FOOTAGE INDICATOR SCALE

CONVENIENT MARKING TEXT
To identify either end of tubing when several single tubes are used together, the numerals 1-30 are printed the full length of the tubing. By cutting off the lead end at a given number, and marking the reel or carton by the same number, you can easily identify the tube at either end for proper connection.

Figure 1-20. Flexible instrument tubing. (*Courtesy of Eaton Corp., Industrial Polymer Products Div.*)

Figure 1-21. Armored tubing with inner jacket. (*Courtesy of Eaton Corp., Industrial Polymer Products Div.*)

1002 armored metal tubing with inner jacketed multiple tube bundle, and Figure 1-22 shows Type 1003 armored metal tubing with outer jacketed multiple tube bundle.

STAINLESS STEEL STICK TUBING

The most common example of poor materials handling practices is damage to the small sizes of stick tubing after it

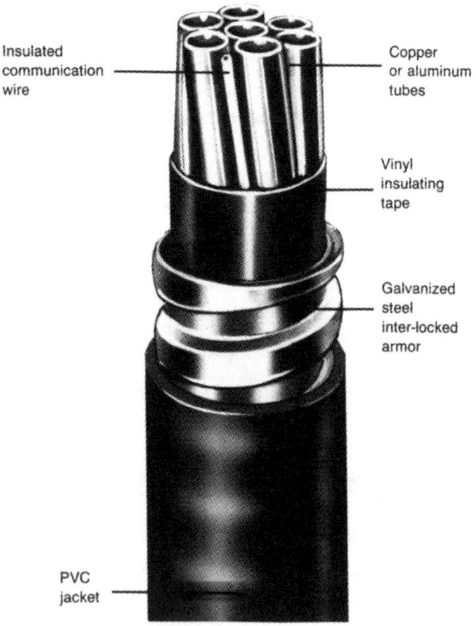

Figure 1-22. Armored tubing with outer jacket. (*Courtesy of Eaton Corp., Industrial Polymer Products Div.*)

has arrived at the project. The tubing is limber, and will require support during storage or while being handled. Always supervise the handling of stick tubing. Figure 1-23 shows several styles of stainless steel stick tubing and cut nipples that are available. The tubing is shipped in 20-ft

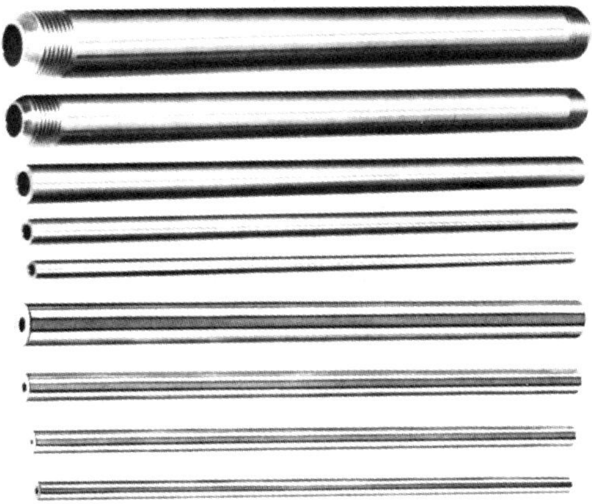

Figure 1-23. Stainless steel stick tubing. (*Courtesy of Autoclave Engineers, Inc.*)

lengths in cardboard tubes, or bundled along a board. It should be stored on a flat surface, or on a rack with support braces every four feet. Tubing is identified by the outside diameter and by the wall thickness. An example would be for stainless steel tubing, ⅜-in. OD by .035-in. wall thickness. If the shipping carton is suitable, leave the tubing in the carton while in the project warehouse to protect the stainless steel finish. Fittings for stainless steel and other instrument items are shown in Chapter 3.

MICROPROCESSOR ELECTRONICS

Control rooms for large industrial plants are based on pneumatics or analog electronics, and have a maze of meters and levers spaced across three or four walls. The miles of wire or tubing in a plant are expensive to install and maintain. Behind the control panels are thousands of feet of tubing and wire that connect the gauges to each other and often to a computer.

Microprocessor-based electronics is changing that. Computers, televisions, calculators and much of the controls industry now rely on the silicon chip. The tiny integrated circuit has eliminated the need for miles of wiring, making possible smaller and more powerful computers.

2
Instrument Valves

Uses for Valves

Primitive man used a plug in a bamboo pipe as a valve to serve his needs. Modern-day valves are used to control the flow of liquids or gases by on-off service, throttling service, and backflow prevention. According to the Walworth Company, about 50% of the industrial valves used fall in the on-off category, 40% in the throttling category, and 10% in the backflow prevention category.

Basically, the six major valve types, which have almost innumerable variations, are as follows:

- ☐ Gate
- ☐ Globe
- ☐ Check
- ☐ Ball
- ☐ Plug
- ☐ Butterfly

Gate Valves

Figure 2-1 illustrates a flanged end gate valve and the flow characteristics. Gate valves are used for full on and off service.

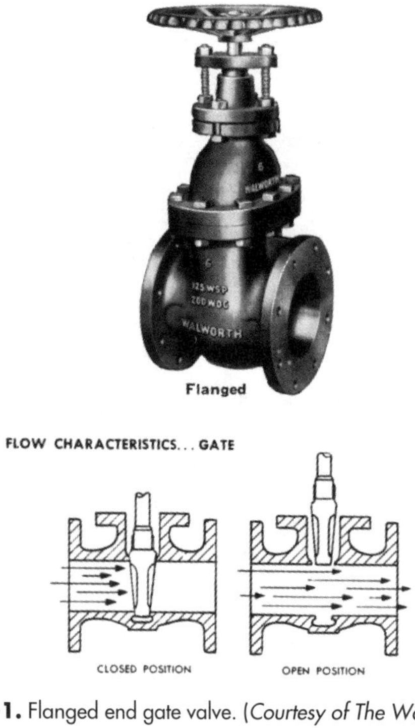

Figure 2-1. Flanged end gate valve. (*Courtesy of The Walworth Co.*)

Globe Valves

Figure 2-2 shows a globe valve with flanged ends and screwed ends. It is sometimes referred to as an "S" valve, and is fast closing. Globes are used mostly for throttling service.

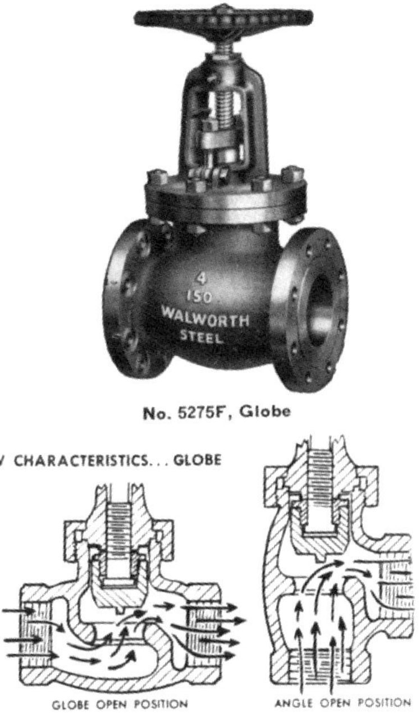

Figure 2-2. Flanged and screwed end globe valve. (*Courtesy of The Walworth Co.*)

Check Valves

Figure 2-3 illustrates a swing check and a globe check with flanged ends. The three basic types of check valves are the swing, lift, and ball. Check valves are designed to prevent backflow.

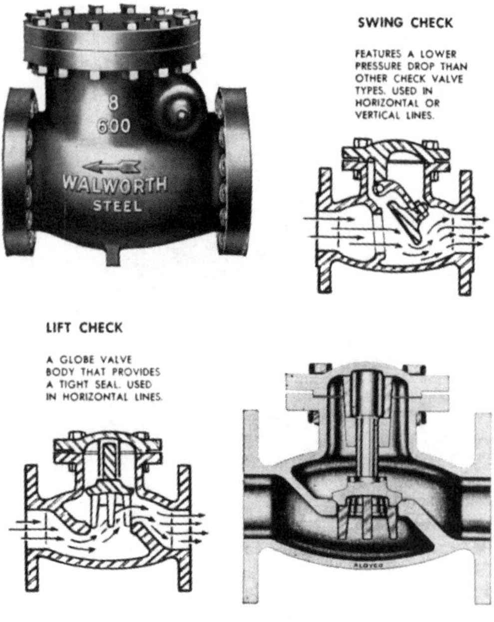

Figure 2-3. Flanged end check valves. (*Courtesy of The Walworth Co.*)

Ball Valves

Figure 2-4 shows three basic styles of ball valves with flanged and screwed ends; full port, reduced port, and the venturi style. Ball valves do not require lubrication and are quick open and quick off.

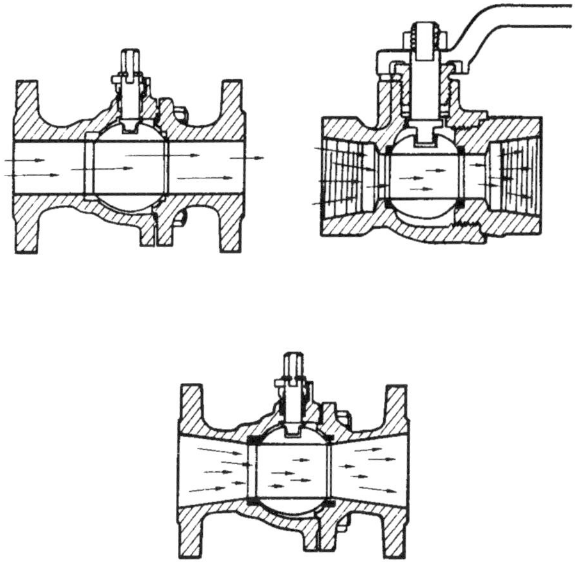

Figure 2-4. Flanged and screwed end ball valves. (*Courtesy of The Walworth Co.*)

Plug Valves

This type of valve has evolved from a tapered plug in a bamboo line to modern versions such as the one shown in Figure 2-5. Lubrication not only ensures ease of operation, but acts to ensure tight sealing and resistance to corrosive solutions.

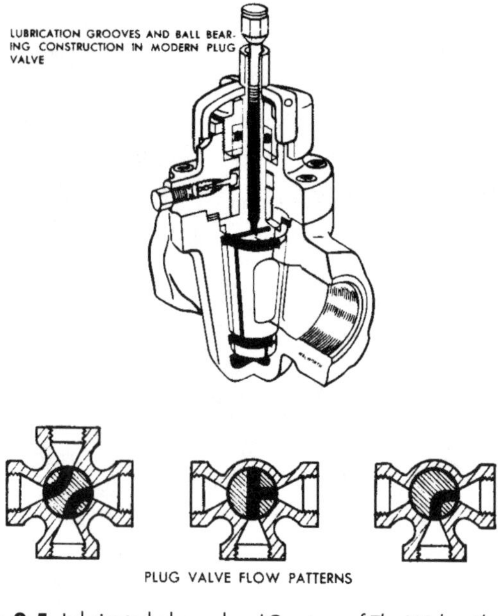

Figure 2-5. Lubricated plug valve. (*Courtesy of The Walworth Co.*)

Butterfly Valves

Figure 2-6 depicts two basic body styles of butterfly valves; the wafer style and the full-lug. Butterfly valves are mounted between flanges. It is the most popular throttling device. More detail on butterfly valves follows later in this chapter.

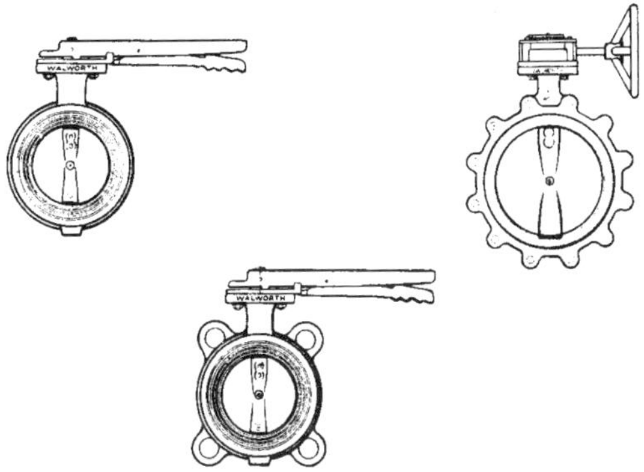

Figure 2-6. Butterfly valves. (*Courtesy of The Walworth Co.*)

Stem Variations

Figure 2-7 illustrates the various stem variations. The descriptions of the variations are important as they will appear in standard valve terminology. An example is OS&Y (outside screw and yoke).

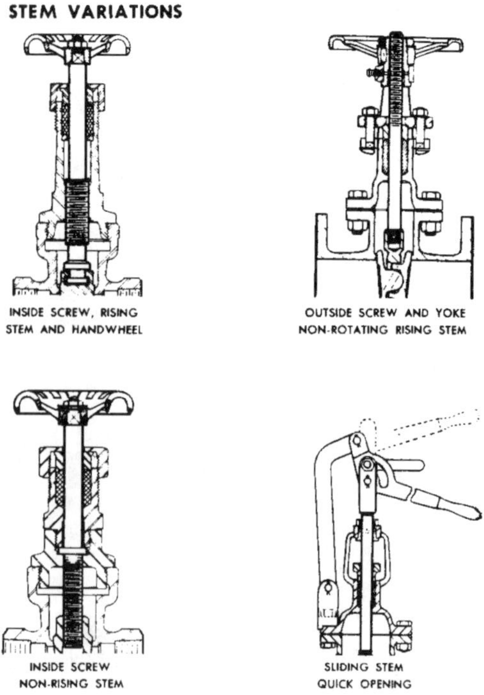

Figure 2-7. Valve stem variations. (*Courtesy of The Walworth Co.*)

Bonnet Variations

The bonnet seals the packing glands around the valve stem. Figure 2-8 shows several styles of bonnets for valves.

BONNET VARIATIONS

Figure 2-8. Valve bonnet variations. (*Courtesy of The Walworth Co.*)

Discs Used in Valves

The disc, when closed onto the valve seats, forms a seal against the flow. Figure 2-9 illustrates how various discs are used in gate valves.

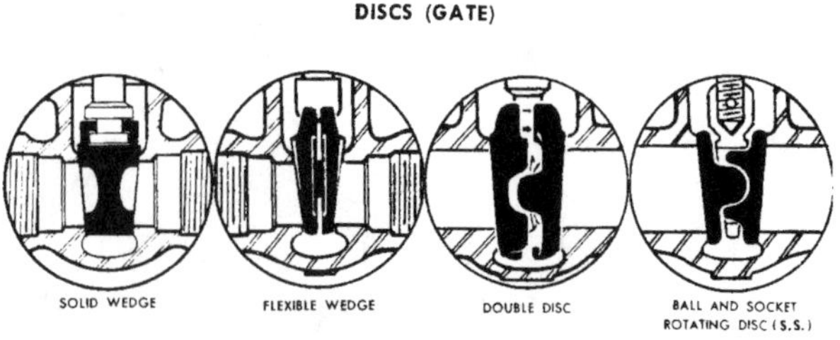

Figure 2-9. End connections for valves. (*Courtesy of The Walworth Co.*)

End Connections for Valves

A few of the more common end connections used on valves are shown in Figure 2-10. There are other variations used by different manufacturers in special applications.

VALVE DESCRIPTION

Describing valves for entry into the materials accounting system is not difficult, but their description must remain consistent with the project commodity code descriptions

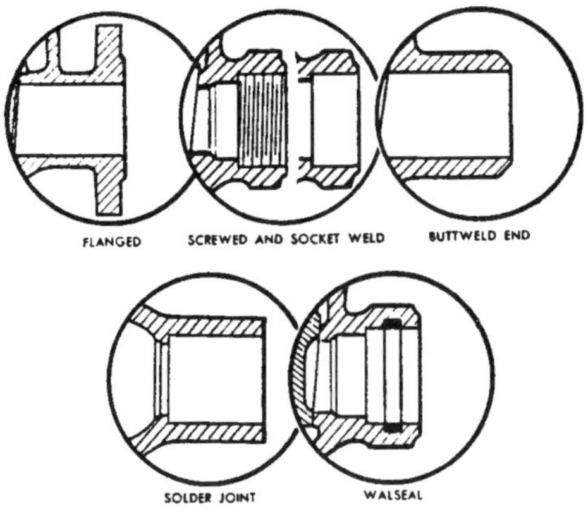

Figure 2-10. End connections for valves. (*Courtesy of The Walworth Co.*)

from the computer bank used for the project. On each valve in one form or another, the manufacturer's nameplate has a valve figure number on it. With this number and the manufacturer's catalog, a good description is available. Basically, the following describe a valve for future use:

1. Size (nominal pipe size).
2. Body style—cast steel, bronze, stainless steel, etc.
3. Stem or ball—11–13.5% chrome, buna-n, viton, etc.
4. Seat materials—reinforced teflon, stellite, 304 stainless steel, metal to metal, etc.

5. Pressure class—150, 300, 600, 900, 1500, 2500 ANSI, 10,000-lb psi, etc.
6. Type of operation—handwheel, lever, worm gear, power operated, chain wheel operator, etc.

Project Tags

Project tags, usually called V-numbers, are unique for the one project only. An example would be V-2237 for a 4-in. Walworth Figure number 5202F cast steel gate valve. The craftsman would simply request from you, a 4-in. V-2237. On the next project, V-2237 may refer to a completely different item. Tags for V-numbers vary from round discs and wire to permanent weatherproof tags as shown in Figure 2-11. These permanent tags, also called dense metal weld identification tags, are made from copper plates ½ × 2½ in. long. They are attached by using a fast-acting bonding agent (Loctite 326), which is included with the tags. The cryogenic and high-temperature dense metal weld identification tags referred to as Models 300 and 400 are made from copper or stainless steel, and furnished with banding and snap-clips. These tags are impervious to corrosion and weather, but most important, will withstand the stress of cleaning machines, such as a sand blasting unit.

Important Rule When Receiving or Issuing Valves

Never remove project tags from valves, and if a tag is off, request your supervisor to verify a replacement tag as

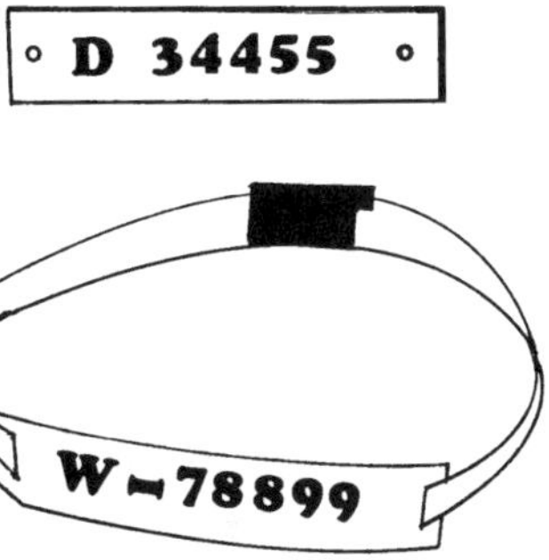

Figure 2-11. Dense metal weld identification tags. (*Courtesy of South Manufacturing Co.*)

the correct number. Do not rely on the project tags. As the responsible materials person, verify the valve as correct by the figure number on the nameplate on each and every valve when either receiving or issuing the valves. Disaster can result from an incorrect valve being installed in a process line. Do not paint valves prior to issue. Surplus valves that have been painted cannot be returned for credit.

Storing Valves

Figure 2-12 shows one method of storing valves on pallets and inside a warehouse under ideal conditions. Valves two inches and smaller, and any valves with threaded outlets, should be stored inside. Valves stored outside must be stored with the stem up, and in such a manner as not to trap

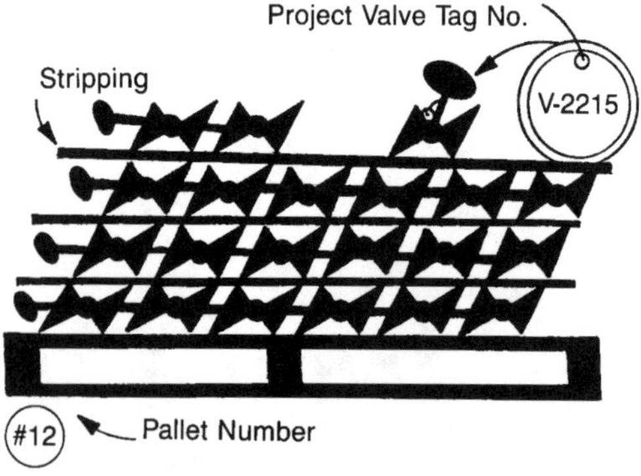

Figure 2-12. Gate valves stored on pallets.

water or sand. Corrosion will form on the ball or seat, and render a new valve worthless even before it is installed in the process line. Use end protectors for a reliable seal against the elements.

Operators for Valves

Hand wheels, levers, and throttling levers are the most common types of devices to operate valves. A chain wheel operator is used to operate hand wheels that are out of reach of the plant personnel (see Figure 2-13).

Butterfly valves are used as control valves. Three styles of operators are shown in Figure 2-14:

When ordering, give diameter of valve wheel to be fitted and No. of feet of chain required.

No.	Diameter Sprocket Wheel Inches	Will Fit Valve Wheels, Diameter Inches
0	4	2 to 4
1	5⅞	4⅛ to 5⅞
1½	7½	6 to 7½
2	9	7¾ to 9
2½	12½	9¼ to 12½
3	15½	12¾ to 15½
3½	19	15¾ to 19
4	22	19¼ to 22
4½	26	22¼ to 26
5	30	26¼ to 30

Figure 2-13. Chain wheel operator. (*Courtesy of The Walworth Co.*)

Figure 2-14. KLOK® valve operators. (*Courtesy of Keystone International, Inc.*)

A. K-Lok valve with electric actuator, with manual override.
B. K-Lok valve with notch plate handle for manual operation/throttling.
C. K-Lok valve with compact pneumatic rotary actuator. (See Figure 2-15 for a cutaway view of the Keystone Figure 790 actuator.)

Figure 2-16 shows three additional types of instrument control valves:

A. 3-way Fisher mixing valve with pneumatic diaphragm actuator.
B. Fisher butterfly control valve, with piston operator, pneumatic positioner for throttling service.
C. Fisher control valve, pneumatic diaphragm actuator.

Valves for Instrument Outlets

All instrument outlets for pressure measurement, vents, drains, and sampling applications require a valve that will withstand the pressure and strain of the process line. Figure 2-17 illustrates a Vogt Class 800 gate valve with an extended end, which has more strength than the traditional valve and nipple concept. (A nipple of incorrect rating installed in a process line could pove to be disastrous.) Applications of

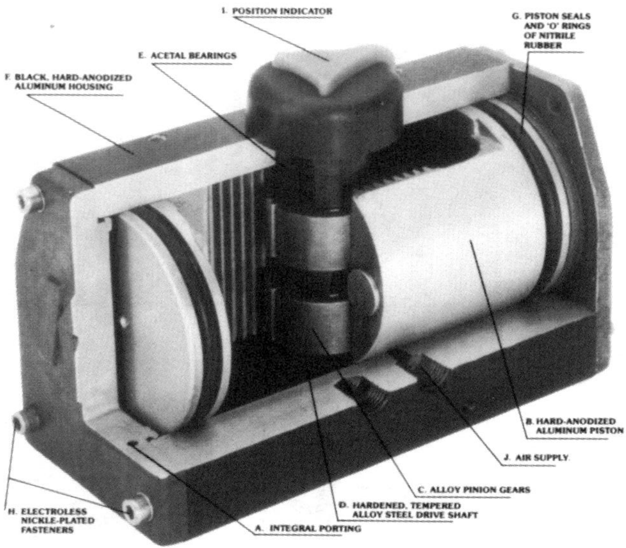

Figure 2-15. Cutaway view of pneumatic valve actuator. (*Courtesy of Keystone International, Inc.*)

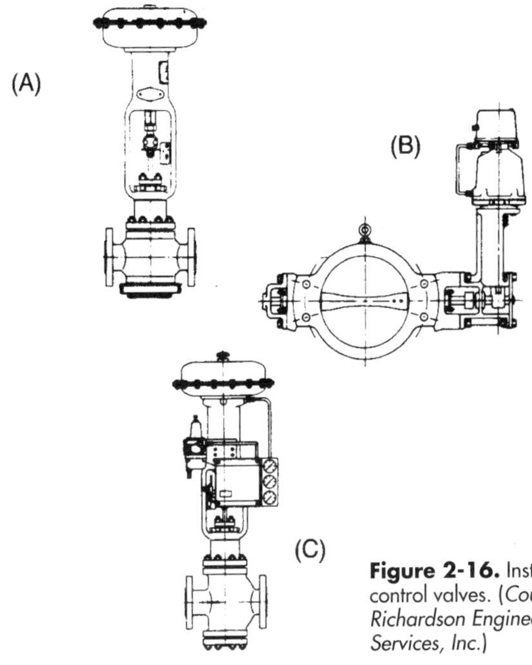

Figure 2-16. Instrument control valves. (*Courtesy of Richardson Engineering Services, Inc.*)

these valves through insulation are shown in Figure 2-18. Insulation is typical and common in most processing plants.

Substitute Valves

Line blinds are used in place of valves to close a line in the process system. There are many variations, and the line

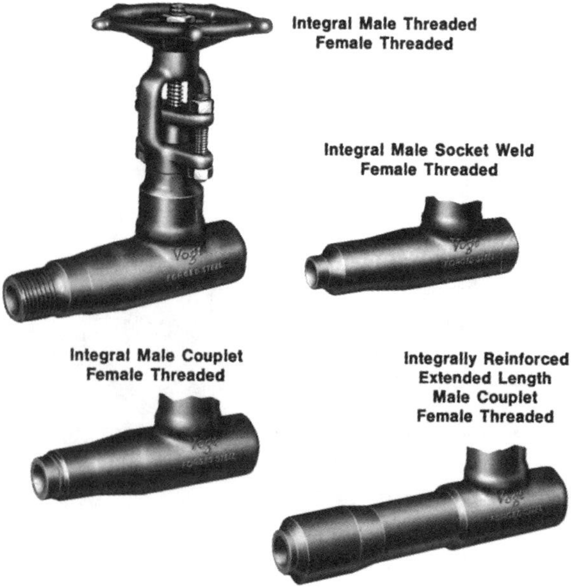

Figure 2-17. Vogt Class 800 gate valve with extended end. (*Courtesy of Henry Vogt Machine Co.*)

blinds are manufactured in most steel types, ANSI pressure classes, and buttweld or flanged ends. They provide a tight shut-off. A Stacey line blind is shown in Figure 2-19 with buttweld ends.

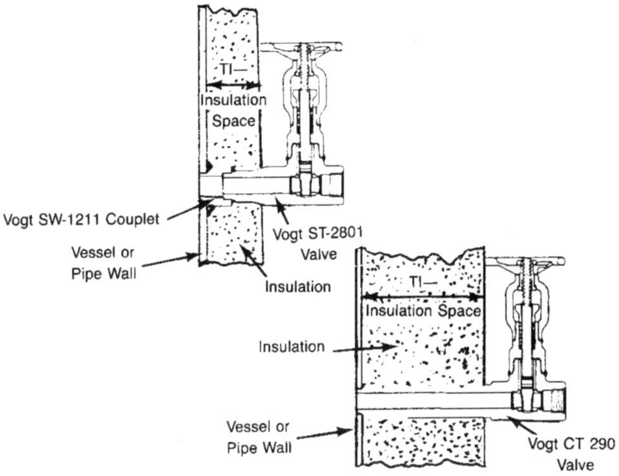

Figure 2-18. Outlet valves through insulation. (*Courtesy of Henry Vogt Machine Co. © Markad Service Co., all rights reserved.*)

Relief Valves

These valves are often referred to as "PSV" valves, and are available in numerous sizes and designs. Figure 2-20 illustrates a typical relief valve in a cutaway view. The spring is designed to release the valve seat at a certain pressure. Earlier models used weights on a long extended handle. Most of the relief valves will be preset at the factory and have a car-seal attached showing the set pressure. Do not

Figure 2-19. Stacey line blind. (*Courtesy of Stacey Fetterolf Corp.*)

remove this seal. In addition, each valve will be tagged on the nameplate for a specific location in the process system.

Relief valves have an outlet that is usually larger than the inlet, and are piped to the vent system or vented to the atmosphere. A full-opening valve is installed below the relief valve, and between the valve and the relief valve, a test insert ring with a gate valve, such as those in Figure 2-17, will be installed. The test insert, which resembles a thick washer with a threaded hole for the outlet, requires a gasket on each side of the ring, and this in turn requires

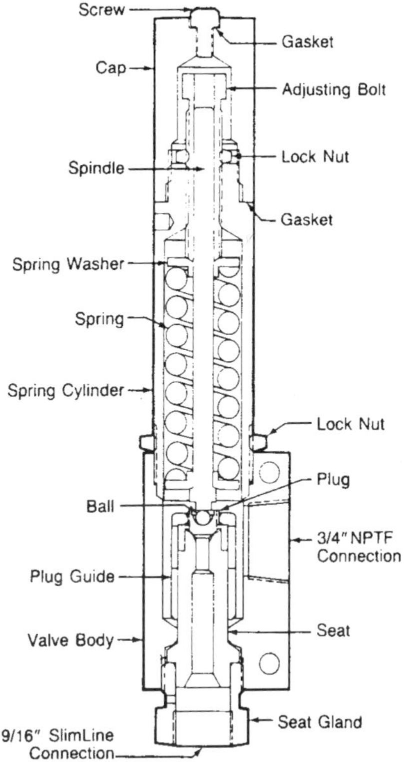

Figure 2-20. Relief valve. (*Courtesy of Autoclave Engineers, Inc.*)

longer than usual stud bolts. A barstock relief valve for high-pressure service is illustrated in Figure 2-21.

Rupture Discs

These are available in a broad range of sizes and types, materials, and rupture pressure ratings. The size is measured

SERIES RVP
RELIEF VALVE

Figure 2-21. Relief valve for high-pressure service. (*Courtesy of Autoclave Engineers, Inc.*)

by the diameter as shown in Figure 2-22. The discs serve as a form of a relief valve, and will be tagged for a specific location and rupture pressure. Always leave the discs in their original carton to prevent damage to the bulge and the edges.

Small Instrument Valves

There are valves of every type and description used in instrument type service. Only a few of the valves manufactured by the leading valve companies are shown here. The same rules for identification and storage that apply to

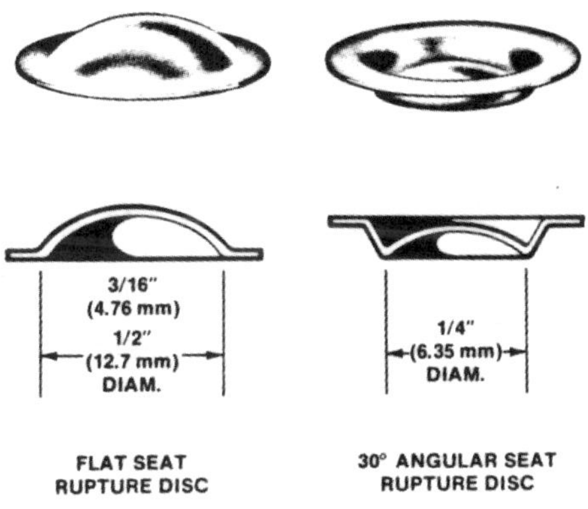

Figure 2-22. Rupture disc. (*Courtesy of Autoclave Engineers, Inc.*)

valves with screwed ends (previously mentioned) apply to these small valves.

Vacuum Valves

Any analytical instrument or industrial process that requires the use of vacuum, such as gas analysis, electronic component manufacturing, cosmetics, freeze dried foods and drugs, requires vacuum valves. Shown in Figure 2-23 is a Nupro® 304-24VFBG Bellows sealed butterfly vacuum

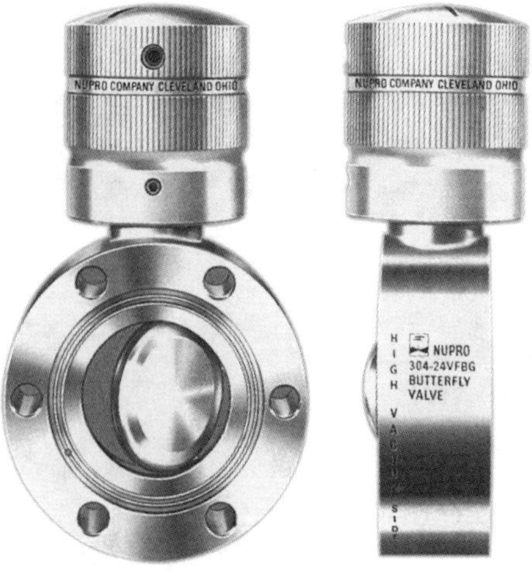

Figure 2-23. Bellows sealed butterfly vacuum valve. (*Courtesy of Nupro® Co.*)

valve. This valve is used in systems with up to 1½-in. tubing, and can be air operated.

Plug Valves

The compact and quick-acting plug valves are used in applications for chemical and petrochemical plants for bleed and drain valves, test equipment, air lines, sampling, and general instrumentation. The valves are color coded for systems to prevent mixing of materials in the process system. They are available in pipe end and tubing connections. Figure 2-24 illustrates plug valves.

Figure 2-24. Instrument plug valves. (*Courtesy of Nupro® Co. © Markad Service Co., all rights reserved.*)

Regulating and Shut-off Valves

Figure 2-25 illustrates a straight-through plug valve used for regulating, and for gauge installation, sampling, and test instruments. The cylinder valve with a rupture disc provides a positive method of over-pressure protection for any system or cylinder. The rupture disc provides instantaneous and unrestricted opening to atmosphere at a predetermined pressure. Figure 2-26 shows a variety of valve patterns and end connections of the regulating and shut-off valves.

Ball Valves for Instrument Service

Small ball valves are used where quick on-off service is required in acid lines, treated water lines, inert gas lines,

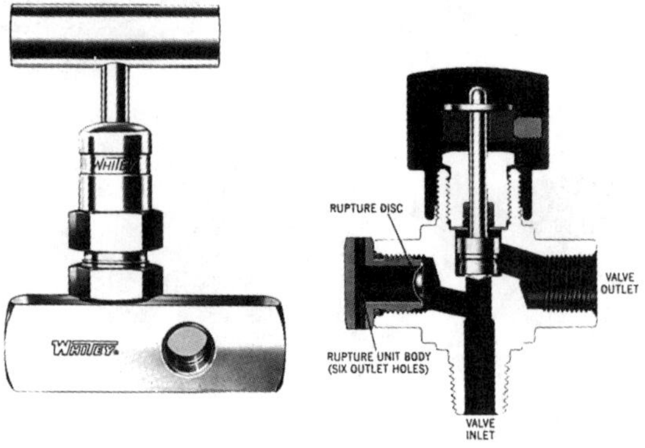

Figure 2-25. Regulating and shut-off valves. (*Courtesy of Whitey Co. ©1972–1978 Markad Service Co., all rights reserved.*)

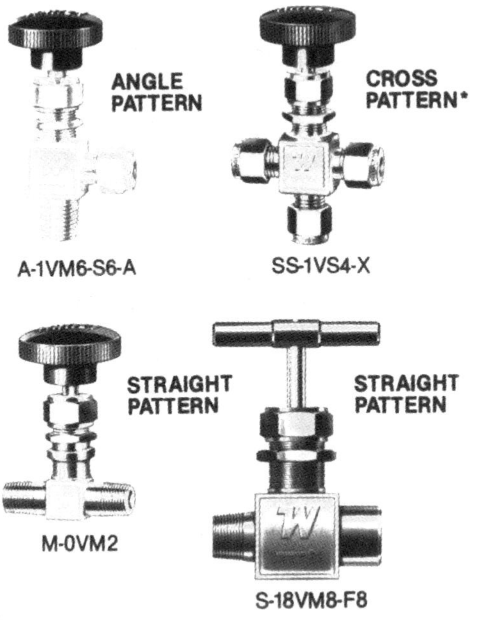

Figure 2-26. Regulating and shut-off valves. (*Courtesy of Whitey Co. © Markad Service Co., all rights reserved.*)

natural gas lines, and gas manifolds. It is used in all types of hydraulic and air lines. Figure 2-27 illustrates a three-way ball valve for switching systems, fluid mixing, sampling and bypass systems (A), a swing-out ball valve that features easy valve maintenance, and offers leak-tight seal-

(text continued on page 54)

(A)

(B)

Figure 2-27. Multiport ball valves. (*Courtesy of Whitey Co. © Markad Service Co., all rights reserved.*)

(C)

(D)

Figure 2-27. Continued.

ing at high- and low-pressures (B), a 5-way ball valve used to direct a number of flows or pressures through a single valve (C), and a ball valve for on and off service (D). These ball valves have a directional handle feature. Figure 2-28 shows a group of fine metering valves of angle and straight

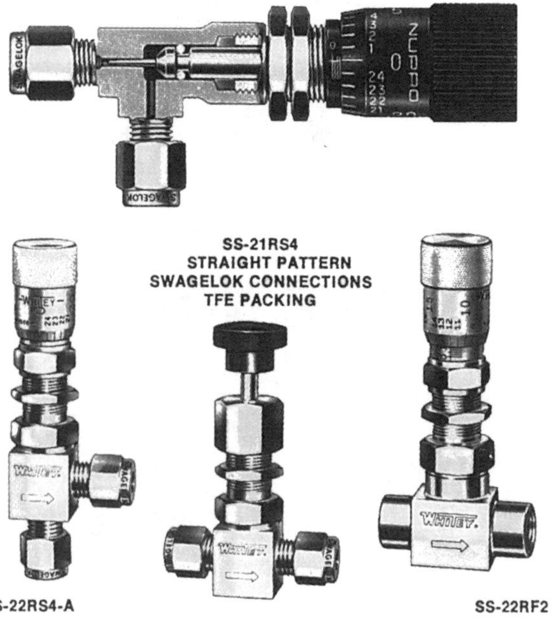

Figure 2-28. Fine metering valves. (*Courtesy of Whitey Co. and Nupro® Co. ©1973–1978 Markad Service Co., all rights reserved.*)

patterns that are used to control gas or liquid flow precisely in critical processes.

Instrument Check, Relief, and Purge Valves

Instrument check valves allow unrestricted flow in one direction, and no flow in the opposite direction. Relief valves are set to open at a preset pressure to protect the gauges, instruments, and systems from over-pressurization. Purge valves are used to manually bleed, vent, or drain instruments and systems. Figure 2-29 shows a variety of these valves with cutaway views: in-line check valve, purge valve, and in-line relief valve.

Valves for High-Pressure Service

Severe service conditions of high pressures up to 150,000 psi, hazardous or corrosive fluids, and very high temperatures to 1,200°F are controlled with valves manufactured by many of the leading valve makers. Figure 2-30 illustrates a valve for service at 45,000 psi at 450°F. These valves, and other makes and patterns, have replaceable port fittings as shown on each side of the illustration. The various fittings can be used to adapt the valves to any tube or piping system.

Valves for extreme pressures are required for various applications such as an offshore oilwell head. A few of such valves are shown in Figure 2-31.

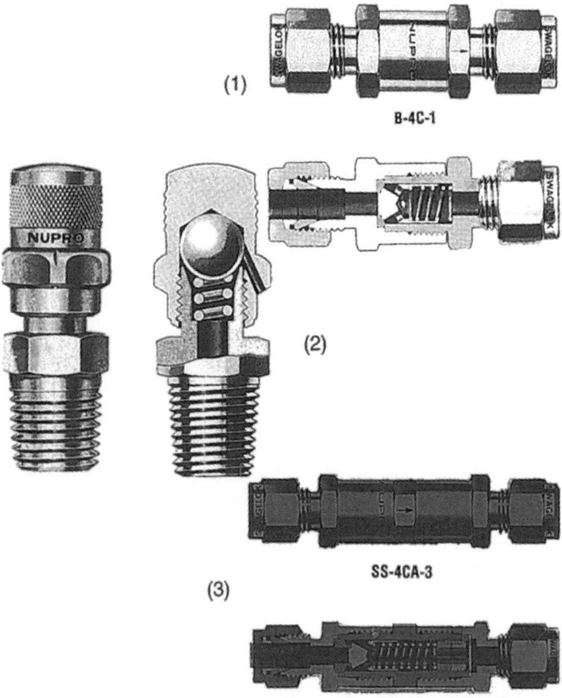

Figure 2-29. Instrument in-line check valve (1); purge valve (2); and in-line relief valve. (*Courtesy of Nupro® Co. ©1973–1978 Markad Service Co., all rights reserved.*)

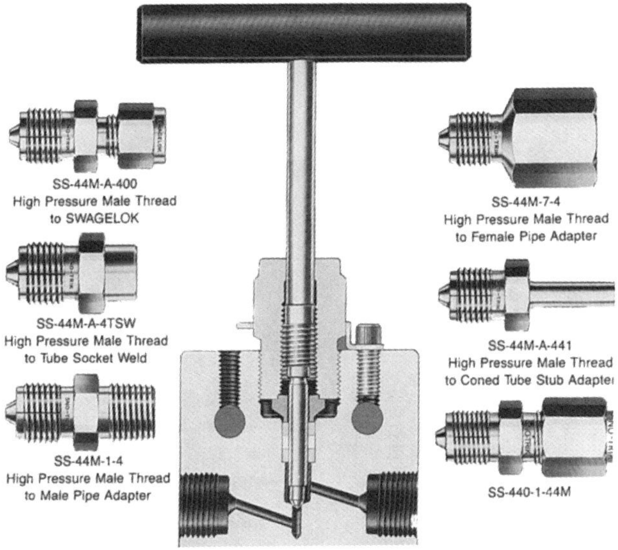

Figure 2-30. High-pressure valve and port adapters. (*Courtesy of Sno-Trik® Co. © Markad Service Co., all rights reserved.*)

Materials Control for Valves

Valves of all types represent a major part of the cost of a project. Regardless of size or ratings, each valve received at your project warehouse has been assigned to serve in a specific line of the process system. It is your responsibility to verify that each valve received or issued complies to the

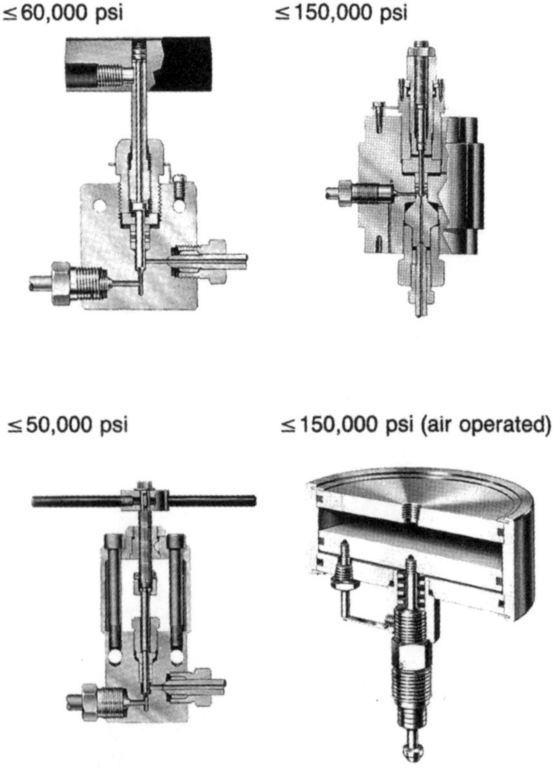

Figure 2-31. High-pressure valves. (*Courtesy of Autoclave Engineers, Inc.*)

purchase order figure numbers stamped on each valve as the manufacturer's model number, and the same number on your project purchase order. If they do not match and the problem cannot be resolved, notify the project manager as soon as possible. Project schedules are usually very tight, and some valves are long-delivery items.

Needle Valves for Accurate Flow

Needle valves are machined from bar stock for reliable performance. These valves are used in oil and gas production, processing plants, or anywhere a flow of fluid must be carefully regulated. Figure 2-32 shows three patterns of needle valves each with two-prong color-coded handles.

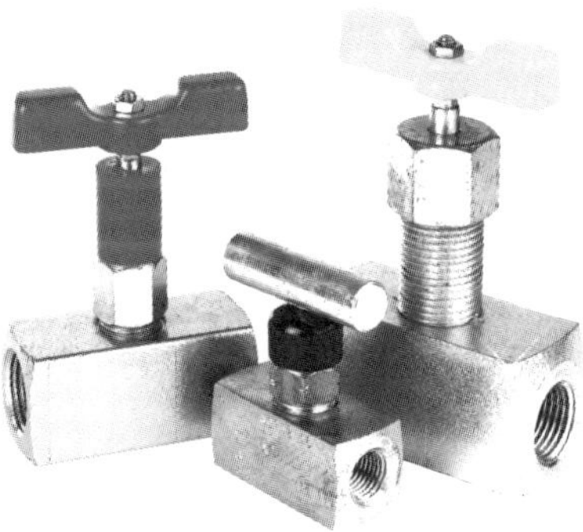

Figure 2-32. Needle valves for accurate flow regulation. *(©Copyright Omega Engineering, Inc. All rights reserved. Reproduced with permission of Omega Engineering, Inc., Stamford, CT 06907.)*

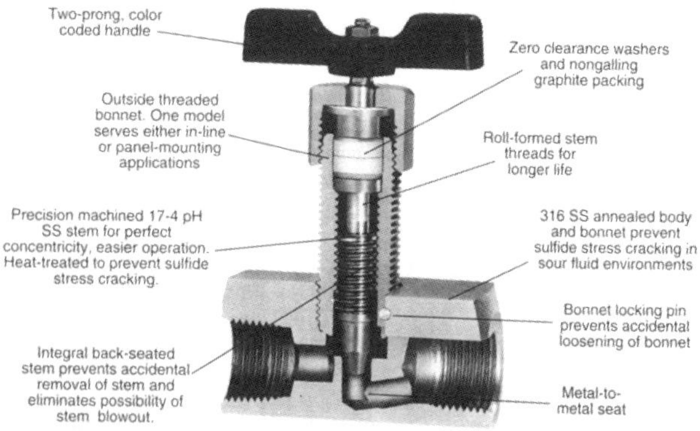

Figure 2-33. Hard seat needle valve for NACE. (*©Copyright Omega Engineering, Inc. All rights reserved. Reproduced with permission of Omega Engineering, Inc., Stamford, CT 06907.*)

Figure 2-33 is a cutaway of a 316 stainless steel needle valve with NACE trim that prevents sulfide cracking per Standard MR-01-75 in offshore and other petroleum environments where sour conditions may be present.

3
Instrument Fittings and Monitors

OUTLET FITTINGS

Outlets on process lines are required near equipment or utility stations for providing plant air or plant gas. There are many types of fittings that can be used for these outlets such as tees, O-lets, and reducing tees. Shown in Figure 3-1 are two outlet fittings called weld couplets. These fittings are acceptable as nozzles on vessels, or can be used on large or small diameter pipe. The threaded couplet is suitable for installing gauges, thermometers, thermocouples, etc.

PIPING FITTINGS

Many standard piping fittings are used in the instrument lines en route to the service area. Figure 3-2 illustrates the basic type fittings whose descriptions remain constant and repetitive. The pressure ratings vary, however, as well as the ends, i.e., screwed, socketweld, combinations, etc.

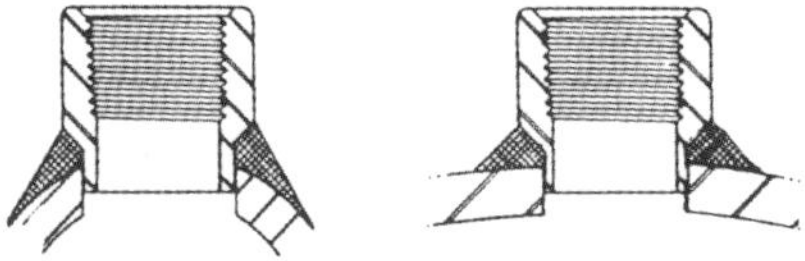

Figure 3-1. Weld couplets. (*Courtesy of Henry Vogt Machine Co.*)

TUBE FITTINGS

Table 3-1 is a cross-reference chart for tube fittings. Part numbers may change without notice. Therefore, it is important to verify the part numbers from the various cata-

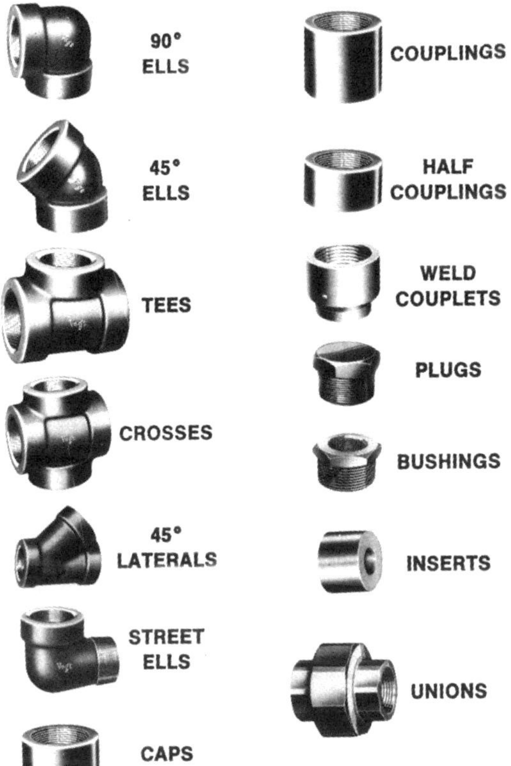

Figure 3-2. Basic types of pipe fittings. (*Courtesy of Henry Vogt Machine Co.*)

logs for project use. The fittings displayed in Table 3-1 are Swagelok®. The part numbers under the other brand names refer to an equal fitting. Regardless of the brand name, tube fittings are made and tested to very high standards to provide service without leaks and other problems. Scratches or other surface defects can lead to gas leaks. Tube fittings should be handled and stored with care to prevent damage. Protect the surface of the fittings by storing them in bin boxes inside their original cartons. (The same rule applies to tubing—never drag it out of a rack or across a cement floor.)

(text continued on page 68)

Table 3-1
Tube Fittings Cross Reference

	Swagelok	Gyrolok	Imperial	Parker	Tylok	
 CAP	1/16 100-C	CP	708-F	FNZ	CAP	
 PLUG	1/16 100-P	P	721-F	PNZ	FPLUG	
 SWAGELOK TO TUBE SOCKET WELD UNION	1/4 400-6-4TWS	—	731-SS	ZHBW	—	
 SWAGELOK TO TUBE SOCKET WELD ELBOW	3/8 600-9-6TSW	—	—	ZEBW	—	MC
 MALE CONNECTOR	1/16 × 1/16-in. 100-1-1	CM	768-F	FBZ		BHMP

MALE CONNECTOR	Swagelok	Gyrolok	Imperial	Parker	Tylok
 BULKHEAD MALE CONNECTOR	1/8 × 1/8 200-11-2	BCM	788-F	FH2BZ	
 MALE ELBOW	1/16 × 1/8 100-2-2	LM	769-F	CBZ	ATPM
 MALE ADAPTER TUBE TO PIPE	1 × 1 16-TA-16		AM	722-F	T2HF
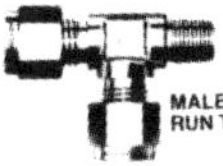 MALE RUN TEE	1/8 × 1/8-in. 200-3TMT	TMT	771-F	RBZ	TMT
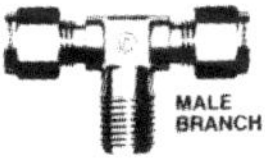 MALE BRANCH TEE	1/8 × 1/8 200-3TTM	TTM	772-F	SBZ	TTM
 FEMALE CONNECTOR	1/8 × 1/8 200-7-2	CF	766-F	GBZ	FC
 BULKHEAD FEMALE CONNECTOR	1/8 × 1/8 200-71-2	BCF	786-F	GH2BZ	BHFP
 FEMALE ELBOW	1/8 × 1/8-in. 200-8-2	LF	770-F	DBZ	FE
 FEMALE ADAPTER TUBE TO PIPE	1/8 × 1/8 2-TA-7-2RT	AF	723-F	T2HG	ATPF

(*table continued on next page*)

Table 3-1 (*continued*)

	Swagelok	Gyrolok	Imperial	Parker	Tylok
 FEMALE RUN TEE	⅛ × ⅛ 200-3TFT	TFT	767-F	MBZ	TFT
 FEMALE BRANCH TEE	⅛× ⅛ 200-3TTF	TTF	777-F	OBZ	TTF
 UNION	⅛-in. 200-6	U	762-F	HBZ	IU
 BULKHEAD UNION	⅛ 200-61	BU	782-F	WBZ	BHU
 REDUCING UNION	5⁄16 × ¼ 500-6-4	RU 756-F	—		RU
 UNION ELBOW	1⁄16 100-9	LU	765-F	EBZ	ELU
 UNION TEE	3⁄16 300-3	TTT	764-F	JBZ	TTT
 UNION CROSS	¼ 400-4	C	752-F	KBZ	CR
 SWAGELOK TO AN UNION	⅛ × ¼ 800-6-4AN	UAN	792-F	XHBZ	—
 PORT CONNECTOR	½ 811-PC-4/6	—	—	2PC	—

	Swagelok	Gyrolok	Imperial	Parker	Tylok
 SWAGELOK TO MALE PIPE WELD CONNECTOR	¼ × ¼-in. 400-1-4MPW	CW	729-FSS	2HBW	—
 SWAGELOK TO MALE PIPE WELD ELBOW	½ × ½ 810-2-8MPW	LBW	—	ZEBW2	
 O-SEAL MALE CONNECTOR PIPE THREAD	¼ 400-1-40R	COM	—	ZHBF5	
 O-SEAL STRAIGHT THREAD CONNECTOR	3/16 × ⅜ 300-1-OR	COS	—	T2HOA5	—
 NUT	¼-in. 402-1	N	761-F	BZ	N
 KNURLED NUT	¼ 402-1K	KN	—	BZB	—
 BACK FERRULE	¼ 404-1	FR	760-F	—	—
 FRONT FERRULE	¼ 403-1	FF	760-F	TZ	FCRC
 SWAGELOK TO AN BULKHEAD UNION	⅛ 200-61-2AN	BUAN	733-F	—	—
 BULKHEAD REDUCER	¼ 400-A1-4	BA	785-F	T2H2BZ	BHA

(*table continued on next page*)

Table 3-1 (*continued*)

	Swagelok	Gyrolok	Imperial	Parker	Tylok
 REDUCER	$\frac{1}{8} \times \frac{3}{16}$ 200-R-3	R	783-F	TRBZ	RATT
 SWAGELOK TO AN ADAPTER	$\frac{1}{8} \times \frac{1}{8}$ 200-A-2ANF	AAN	734-F	X6HBZ6	—

(text continued from page 64)

PARTS DISPLAYS AND MANUALS

The various manufacturers of the tube and tube fittings will be happy to supply your project with poster-type illustrations of their fitting with each component part shown in the illustration. There are manuals for tube fitting and installation that cover topics such as terminology, trouble shooting, types of fittings by number, and other data useful to the craftsmen on the job. Figure 3-3 is an example of the topics that appear in the manuals and catalogs. Request these manuals through your project manager.

HIGH-PRESSURE TUBE FITTINGS

Fittings for extreme pressures are shown in Figure 3-4. A typical use for high-pressure tube fittings and valves is illustrated in Figure 3-5, an oilwell tubing head. The Bourdon gauge (see inset in Figure 3-5) operates from force effect, and uncurls in response to internal pressure in a pro-

1 Simply insert the tubing into the SWAGELOK tube fitting. ***Make sure that the tubing rests firmly on the shoulder of the fitting and that the nut is finger-tight.***

2 Before tightening the SWAGELOK nut, scribe the nut at the 6 o'clock position.

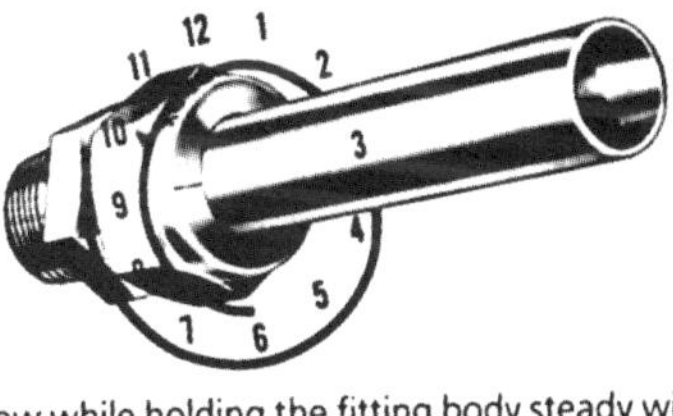

3 Now while holding the fitting body steady with a backup wrench, tighten the nut 1-1/4 turns.* Watch the scribe mark, make one complete revolution and continue to the 9 o'clock position.

Figure 3-3. Installation instructions for tube fittings. (*Courtesy of Crawford Fitting Co.*)

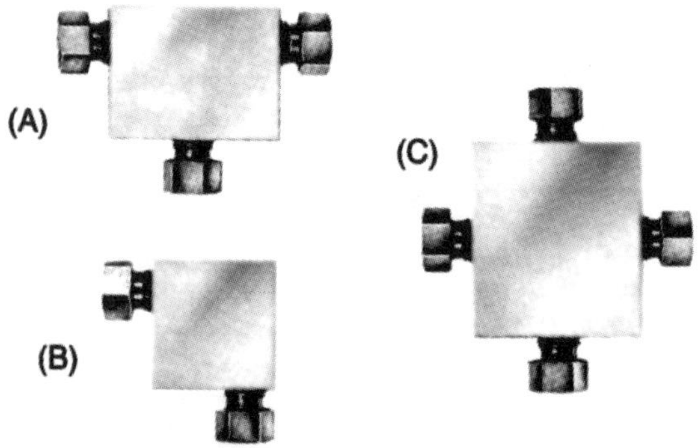

Figure 3-4. High-pressure tube fittings: tubing tee (A); tubing elbow (B); and tubing cross (C). (*Courtesy of Autoclave Engineers, Inc.*)

portional manner. The more sophisticated well heads have shut-down devices, automated flow controls, and are triple completed (or more), within the many strings of pipe. The wells can be controlled at several stages—the test separator, the disposal pit or caisson, on-off flow cycles, etc. If a line breaks, the pressure-drop activates a shut-down valve that closes the line immediately, and then sounds an alarm in the control panel.

IN-LINE FILTERS

Small in-line filters remove contamination as small as two microns. The filter bodies are not removed from the

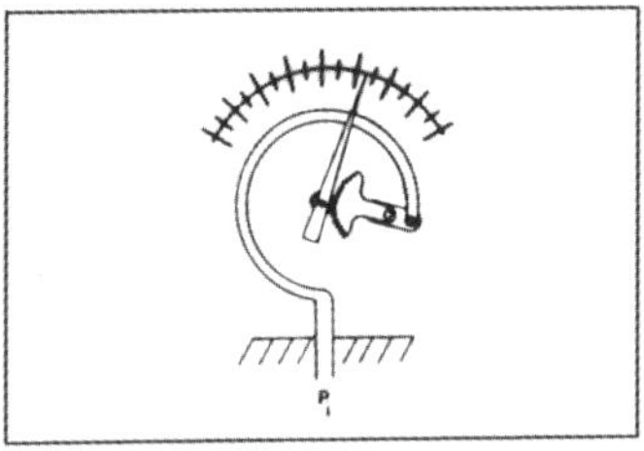

Bourdon gauge. (Courtesy of Transmation, inc.)

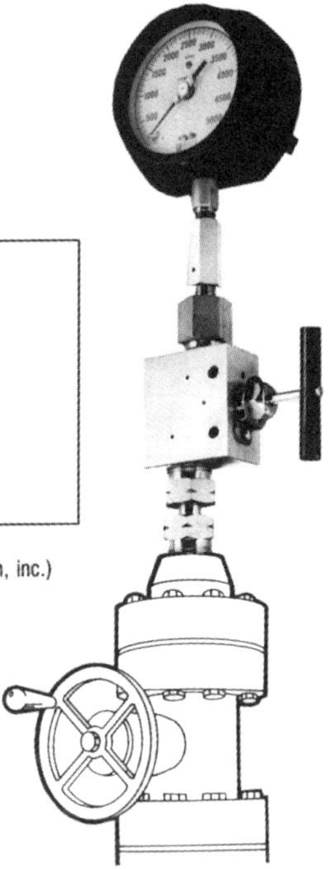

Figure 3-5. Typical oilwell head with valve and pressure gauge. (*Courtesy of Autoclave Engineers, Inc.*)

line to change the elements. For sampling, a bypass feature is provided on some of the filters. There are several sizes of elements used depending on the size of the particles to be removed. Figure 3-6A shows the "TF" series of in-line fil-

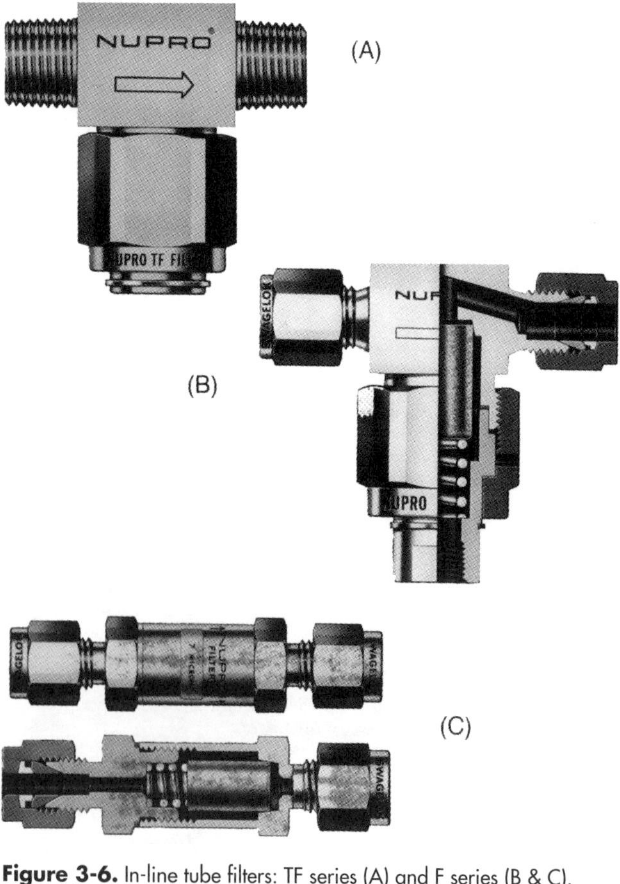

Figure 3-6. In-line tube filters: TF series (A) and F series (B & C). (*Courtesy of Nupro® Co.*)

ters that offers a bypass version and easy in-line removal of the element; Figure 3-6B shows an "F" series, a compact model where space is limited (the element is also replaceable). Figure 3-6C is also an "F" series. Figure 3-7 illustrates two types of elements, a sintered and strainer. As fluid enters the filter, the stainless steel sintered element or strainer will stop the contamination, and permit the clean fluid to pass.

STEAM TRAPS

Lost energy increases the cost of usable energy. Air escaping from under your door during the air conditioning

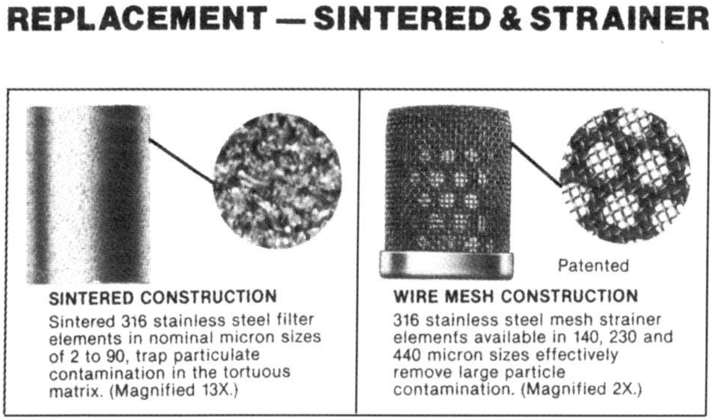

Figure 3-7. Filter elements. (*Courtesy of Nupro® Co. © Markad Service Co., all rights reserved.*)

season will ensure a larger electric bill for you. The same applies to plants that lose money through steam leaks in steam tracing traps, line drip traps, and process traps.

The primary purposes of steam traps on steam line and branch lines are to drain condensate, purge air and non-condensibles to assure dry steam supply to process equipment, and protect the steam piping system from erosion, corrosion, and water hammer. Steam tracing is used to provide a uniform temperature on process lines and equipment in hostile areas where freezing temperatures prevail. One of the most important services provided by traps is to prevent the accumulation of condensate from forming water slugs in lines. These slugs of water cause water hammer, which, when carried through a line at 100 mph, can cause serious damage to control valves, heating coils, and other components. Condensate forming in a line and an example of water hammer are shown in Figure 3-8.

Types of Steam Traps

There are three major types of steam traps:

- *Mechanical steam traps,* shown in Figure 3-9, are also known as "bucket traps." They work by the weight of the bucket, together with changing buoyant forces and the pressure differential across the seat, providing on-off discharge of condensate near steam temperature. Another style, which is furnished with a strainer, is the UniBody™ trap shown in Figure 3-9, bottom.

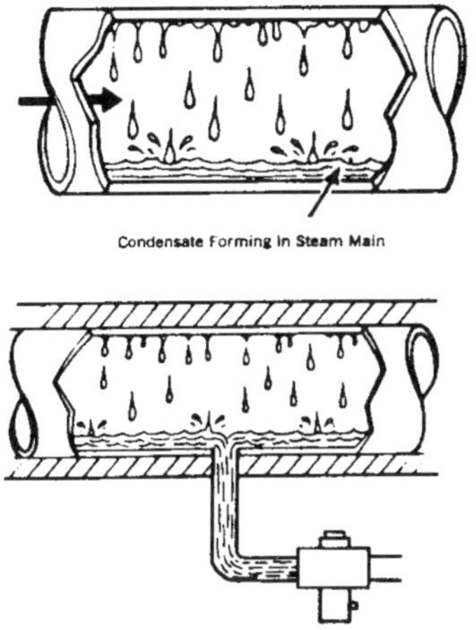

Figure 3-8. Condensate forming in main steam (top) to create water hammer; condensate trap (bottom). (*Courtesy of Yarway Corp.*)

- *Thermostatic traps* are activated by the temperature of the fluid reaching the trap. The thermostatic diaphragm type trap illustrated in Figure 3-10 is actually a complete condensate drainage station including a full-size Y-type strainer and blow-off valve.
- *Thermodynamic traps,* as shown in Figure 3-11, are used with flash steam, which is caused by a system discharging hot condensate into a lower pressure system, developed during discharge of hot condensate to control the opening and closing of the trap.

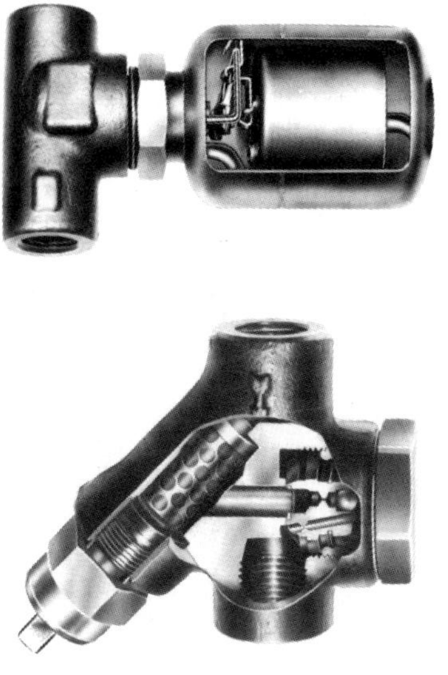

Figure 3-9. Mechanical steam traps. (*Courtesy of Yarway Corp.*)

Four other traps are shown in Figure 3-12 to increase your ability to recognize the various styles of traps:

A. A dual range trap that has two valves—the pilot valve, which functions thermostatically, and the main valve, which stays closed and opens when the pilot valve's load becomes excessive.

B. A bimetallic steam trap for instrumentation and winterization tracings. The operating principle is based on two opposing forces.

Figure 3-10. Thermostatic diaphragm type steam trap. (*Courtesy of Yarway Corp.*)

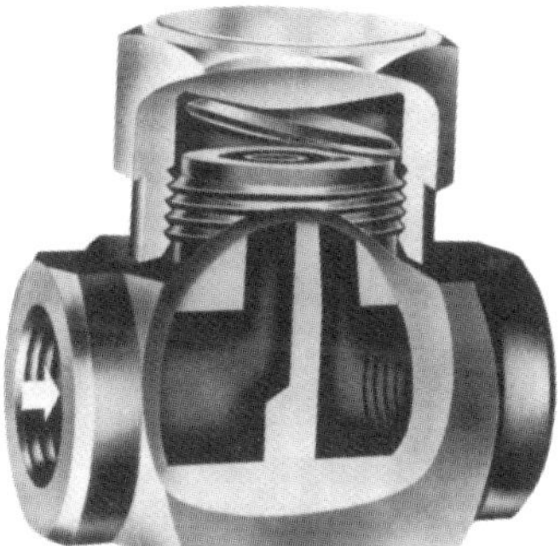

Figure 3-11. Thermodynamic steam trap. (*Courtesy of Yarway Corp.*)

Figure 3-12. Types of steam traps. (*Courtesy of Yarway Corp.*)

C. Another bimetallic trap, which has a bimetallic pilot that controls a large diameter discharge valve, and operates on the principle based on two opposing forces. Used in large reboilers and heat exchangers.

D. Also a bimetallic steam trap that operates similar to B and C.

(C)

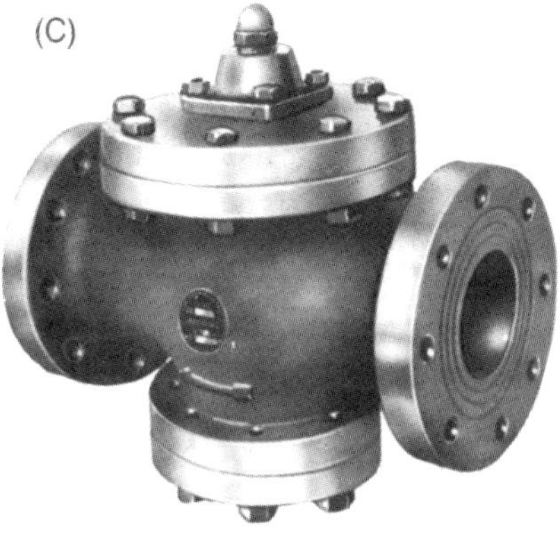

(D)

Figure 3-12. Continued.

STEAM TRACING CONTROL VALVE

It is obvious that operation costs would soar if steam lines used for tracing instruments and process lines had to be left open until some plant operator could shut them down, when the temperature is above freeze conditions. Figure 3-13 shows a temperature responsive ambient con-

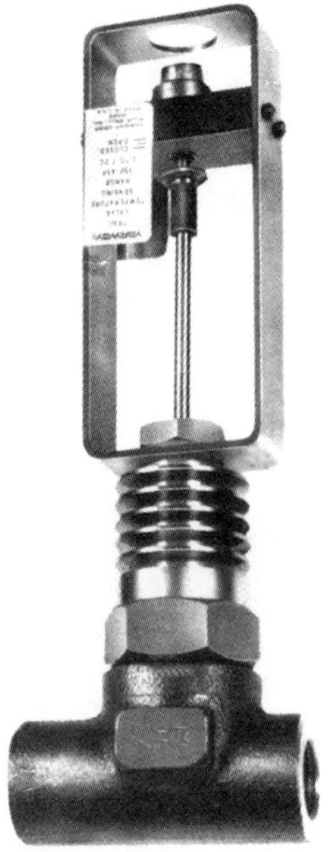

Figure 3-13. TRAC™ valve. (*Courtesy of Yarway Corp.*)

trol valve. The valve automatically opens and closes at preset actuation points. Warm air causes the thermostat to expand, cold air causes it to close. Figure 3-14 shows the manufacturer's recommended installation plan, and is a good example of how the traps, drains, and tracing operate.

SAMPLE CYLINDERS

These are used to safely and reliably contain pressurized fluid samples. They permit the safe handling of fluids during sampling and transporting to the laboratory. The cylin-

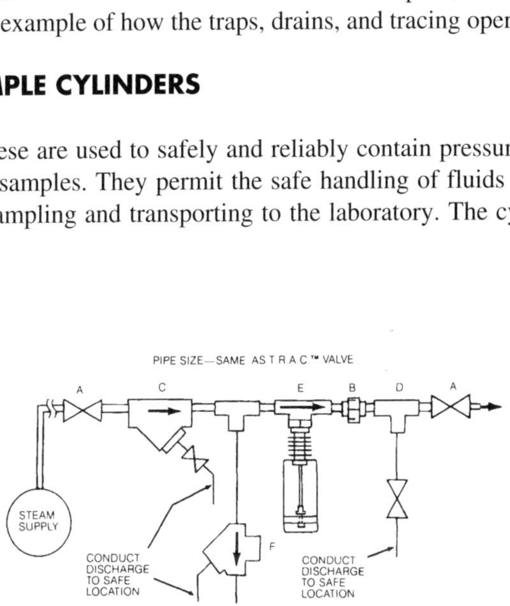

A—Full ported stop valves. End connection not less than T.R.A.C. valve end connection.
B—Unions—installed on downstream side, only if needed.
C—Yarway 901 (cast iron) or 921 (cast steel) strainer.
D—Test tee for inspection of T.R.A.C. valve.
E—T.R.A.C. valve.
F—½" Series 720 or 710 drip trap, with weather cap, to protect the system against condensate accumulation.

Figure 3-14. Piping plan for steam control. (*Courtesy of Yarway Corp.*)

ders are safe for corrosive, toxic, hazardous, or very expensive fluids. They meet applicable D.O.T. Specifications. Four styles of sample cylinders are illustrated in Figure 3-15. Smaller cylinders are also available and used to divide samples into several smaller samples that may be subjected to a series of analyses.

DETECTORS AND MONITORS

Gas leaks out of the process areas can be detected by several means. One method is to use portable units, as

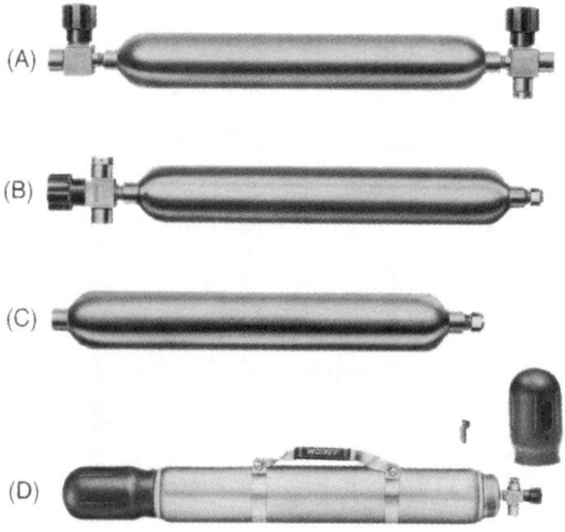

Figure 3-15. Sample cylinders. (*Courtesy of Whitney Co. © Markad Service Co., all rights reserved.*)

shown in Figure 3-16. One unit is equipped with a metal probe, rubber hose, and soft rubber aspirator, which draws a sample from manholes, bar holes, and passes it through the measurement cells. The two-position valve is first set in

Figure 3-16. Combustible gas detectors. (*Courtesy of Southern Cross Corp.*)

the high range after purging with air. If the reading is very low in this range the valve is switched to the low range. No additional adjustment is required when switching from one range to another. The second portable unit is shown in the case on top of the metered unit. The flame pack, as it is called, is a sensitive, single-purpose gas leak detector. It uses the flame ionization principle and patented circuitry. A high-capacity, fully automated air sampling unit is also available, which mounts on the front of a vehicle, and will analyze the entire intake sample.

Bar holes made in asphalt, if not plugged, may later enlarge to pot holes, and erode the plant paving area. Figure 3-17 shows how to patch bar holes made during leak detection periods.

Figure 3-18 shows a pipe and cable detector (left) and a valve box and manhole locator (right). The Pipe Horn™ (left) can be used to trace wiring in buildings, telephone lines, ground wires, etc., as well as pipe and cables. The detector on the right is used in conjunction with the Pipe Horn™ and can detect whether or not the object is directly over the buried pipe or cable. It can also be used to locate cast iron drain lines, conduits, and piping in reinforced concrete slabs or walls.

INTERFACE DETECTORS

A device that detects changes in the sound velocity rather than changes in density can detect different liquids in

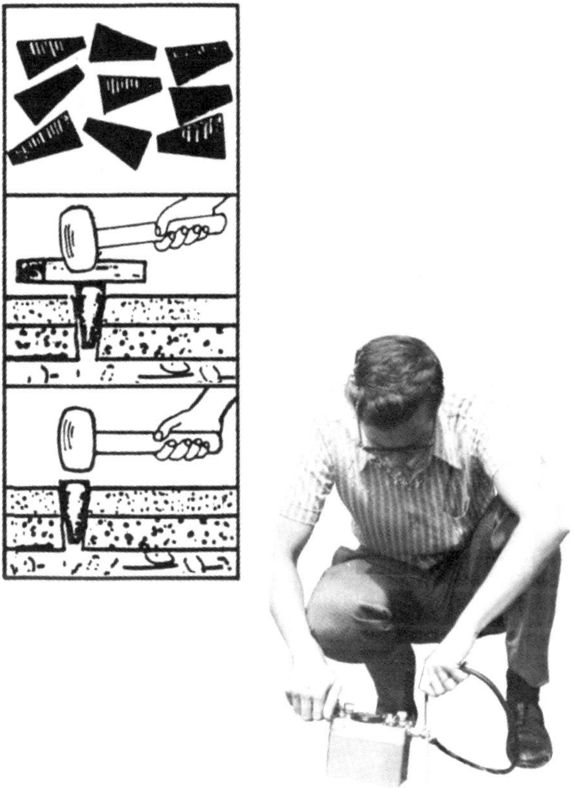

Figure 3-17. Bar hole plugs and sampling device. (*Courtesy of Southern Cross Corp.*)

Figure 3-18. Pipe and cable detectors. (*Courtesy of Southern Cross Corp.*)

a pipeline, even liquids with the same specific gravity such as leaded or unleaded gasoline. Figure 3-19 illustrates the Nusonics® pipeline interface detector that applies microprocessor-based electronics to produce the sound velocity readings for petroleum products. There are no moving parts in the probe assembly.

FLOW SWITCH MONITORS/LEVEL INTERFACE CONTROLLERS

Flow, level, and interface monitors are designed to sense variations in the heat transfer properties of any liquid or

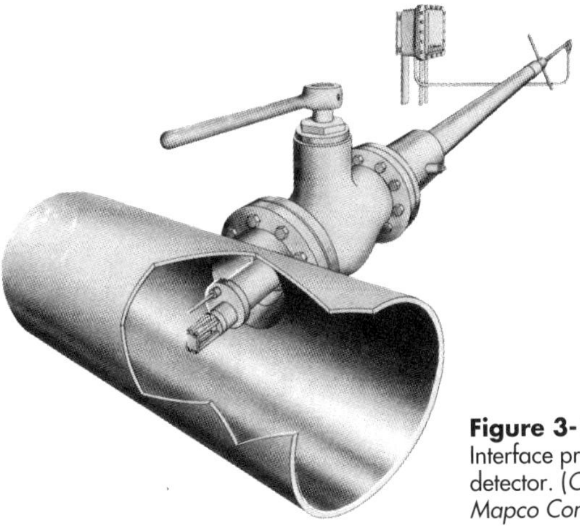

Figure 3-19. Interface products detector. (*Courtesy of Mapco Controls Co.*)

gas. In this manner, reduction or stoppage of flow, loss of product, change in phase from wet to dry and dry to wet, or the interface between dissimilar products can be accurately detected.

Figure 3-20 shows three flow switch monitors: (A) detects reduction or loss of flow in any liquid or gas; (B) responds to a change in flow rate or liquid level; (C) is extremely sensitive to low-flow rates. Similar devices include relief valve leakage monitors, sample rate verifiers, injection rate monitors, bearing lube oil monitors, bearing seal leak detectors, and low- and high-level pump start and stop alarms.

SPARE PARTS

Most of the items described in this chapter, except the pipe fittings, may have spare parts sent to the project for future use. The spare parts are usually on a separate purchase order and tagged as such. These items should be checked and received, left in their original cartons, labeled with project or client commodity code numbers, and then entered into the materials accounting system.

ACTUAL SIZES OF PIPE/TUBING

Learning to instantly recognize pipe and tubing sizes may require considerable practice and skill, and at times

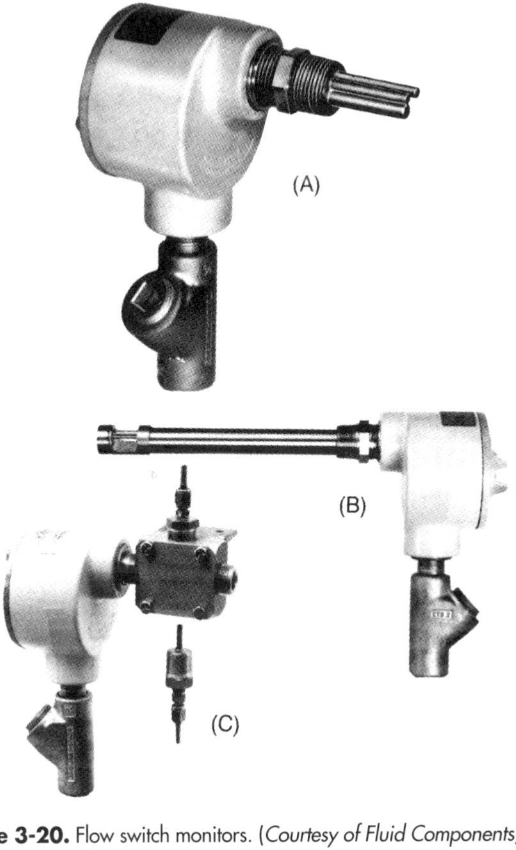

Figure 3-20. Flow switch monitors. (*Courtesy of Fluid Components, Inc.*)

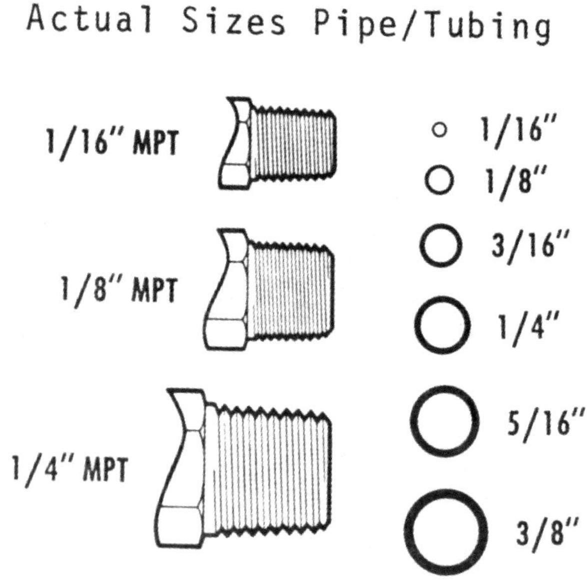

Figure 3-21. Actual sizes of pipe/tubing. (*Courtesy of Crawford Fitting Co.*)

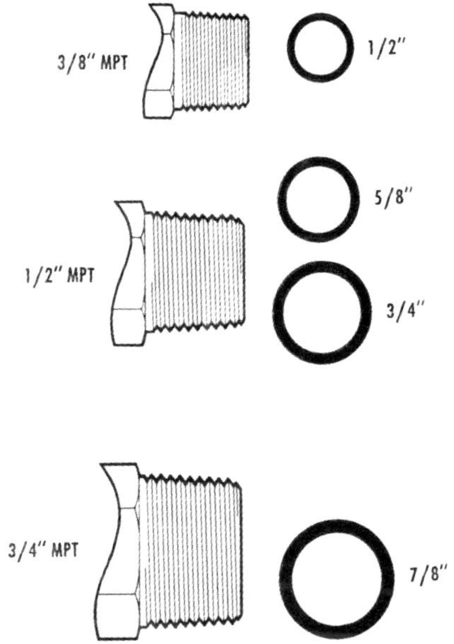

Figure 3-21. Continued.

may cause an incorrect size to be entered on the receiving report. Always measure the size of pipe or tubing, or use a template. Figure 3-21 shows the actual sizes for male nomi-

(text continued on page 95)

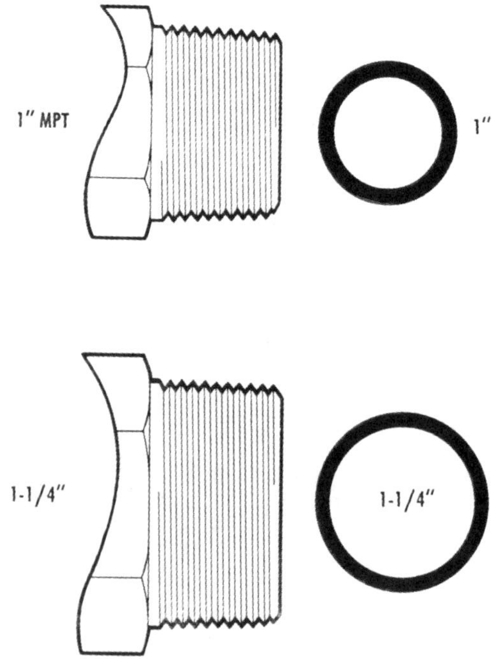

Figure 3-21. Continued.

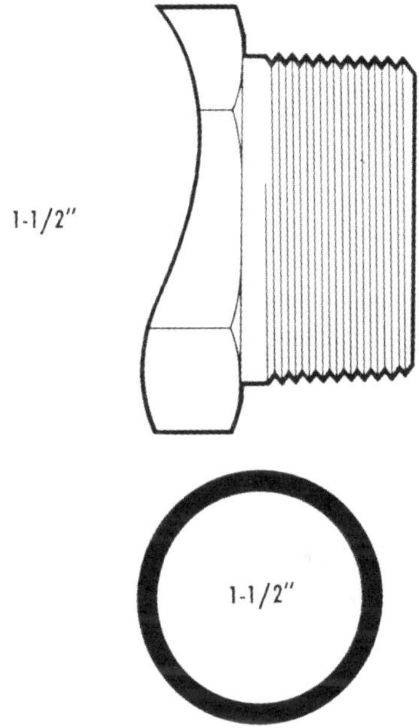

Figure 3-21. Continued.

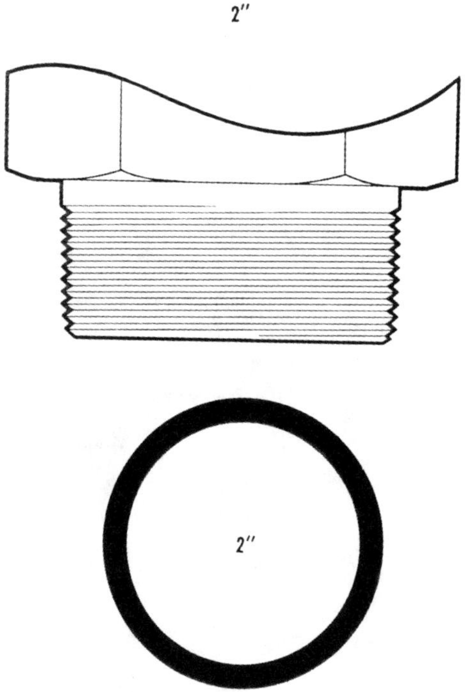

Figure 3-21. Continued.

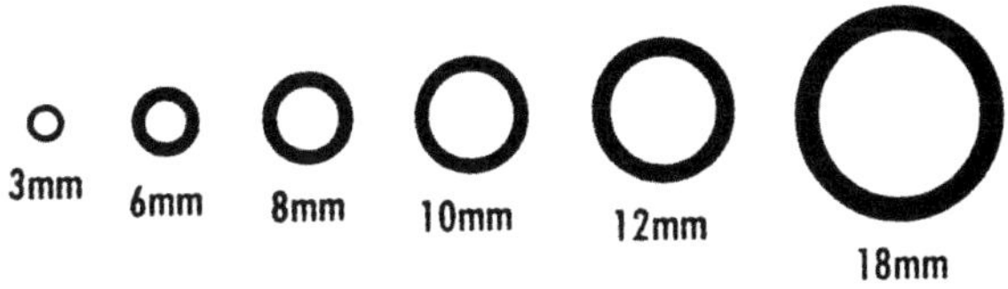

Figure 3-22. Actual sizes of pipe/tubing—metric. (*Courtesy of Crawford Fitting Co.*)

(text continued from page 91)

nal pipe threads and tubing sizes. Figure 3-22 shows the actual sizes of metric tubing through 18mm.

4

Storage Batteries and Solar Modules

AC AND DC SYSTEMS FOR OFFSHORE APPLICATIONS

When normal electrical power fails on an offshore drilling platform or on a production platform, nickel-cadmium block batteries supply the power for engine and turbine starting systems (Figure 4-1) and for emergency lighting and communications (Figure 4-2). This type of battery is also mounted on various configurations of racks, and each rack may contain several batteries.

Receiving the Shipment

Unpack the batteries completely. Do not throw out small packages that may be in the packing materials. The batteries are shipped either filled and charged or empty and discharged. A plastic film is used as a transport seal on the

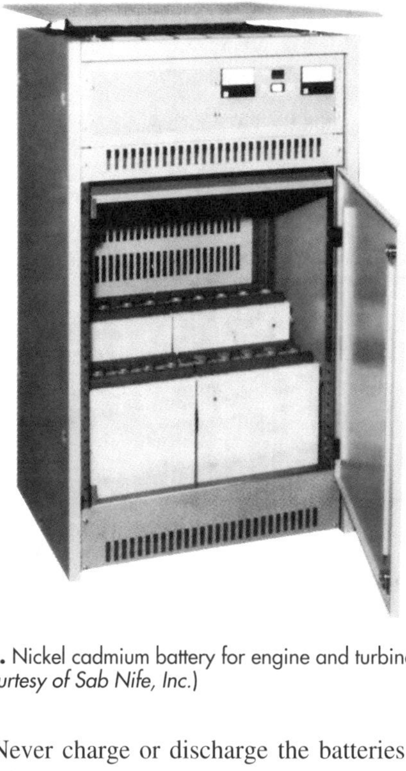

Figure 4-1. Nickel cadmium battery for engine and turbine starting systems. (*Courtesy of Sab Nife, Inc.*)

batteries. Never charge or discharge the batteries until the plastic seal is removed.

Safety Precautions

The nickel-cadmium block batteries are an alkaline system with potassium hydroxide as electrolyte. The alkaline

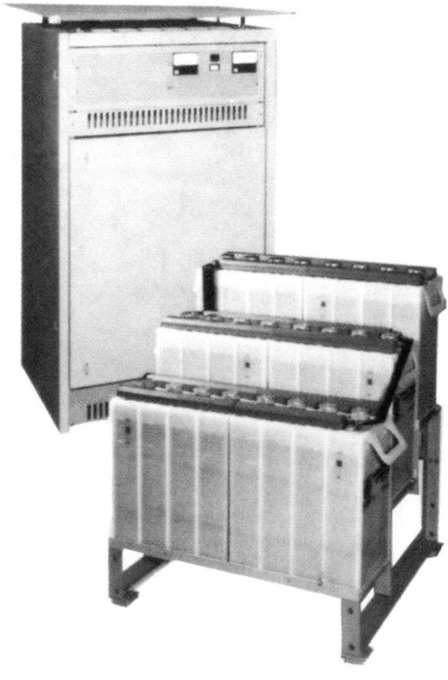

Figure 4-2. Nickel cadmium battery for emergency lighting and communications. (*Courtesy of Sab Nife, Inc.*)

electrolyte is a strong caustic agent, so wear rubber gloves, eye protection and long-sleeved clothing when handling the batteries. Remove rings, watches, and articles of clothing that may come in contact with the battery terminals. The

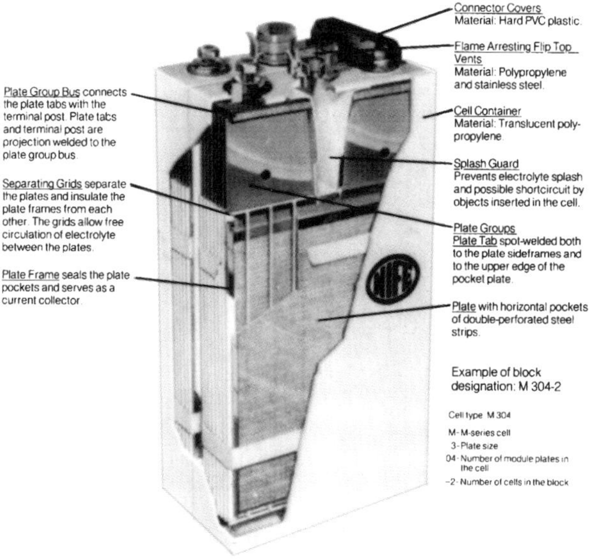

Figure 4-3. Cutaway view of nickel cadmium battery. (*Courtesy of Sab Nife, Inc.*)

terminals are illustrated in a partial view of nickel-cadmium battery in Figure 4-3.

Storing Batteries

Figure 4-4 illustrates several models of the nickel-cadmium block battery that you may store until project startup. Regardless of what kind of battery you may receive, never

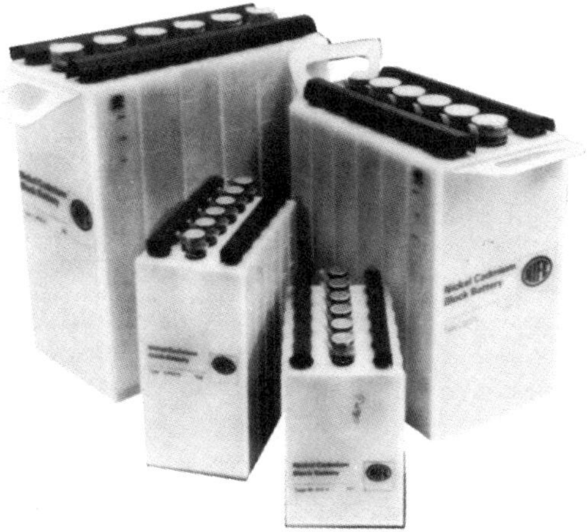

Figure 4-4. Styles of nickel cadmium batteries. (*Courtesy of Sab Nife, Inc.*)

store a battery on a metal deck or on a concrete floor. Always store them on a decking of plywood or boards.

The battery storage room should be dry, clean, and preferably cool. Fully discharged and empty batteries can be stored for an unlimited time, provided the cells are hermetically sealed. A battery filled with electrolyte that is to be stored for a maximum of one year should be charged before storing. No charging is needed during the storage period. However, before being put into service, the battery should be given a normal charge. The battery storage room should be vented. Battery gas is explosive.

PHOTOVOLTAIC SYSTEMS

Figure 4-5 illustrates how the production of electricity starts as soon as sunlight strikes the photovoltaic module. Solar cells within the module will respond to the light by creating an electrical current that moves through the series of cells and is then stored in a battery, such as the one shown in Figure 4-6.

Many photovoltaic systems are used offshore to produce current for warning lights and other components of the power system. A more simple example of how it works is

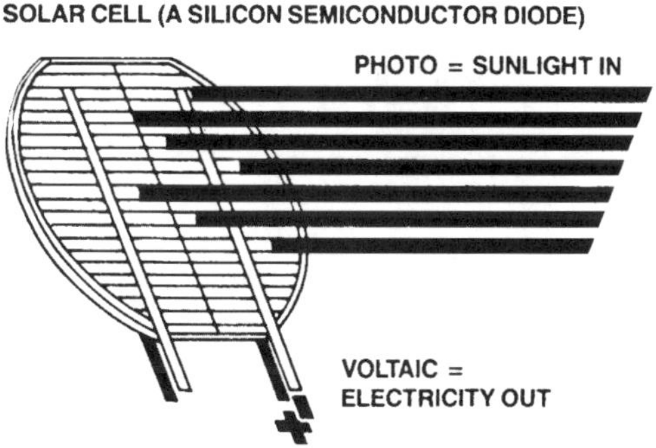

Figure 4-5. Solar cell (silicon semiconductor diode). (*©ARCO Solar, Inc., a subsidiary of Atlantic Richfield Co.*)

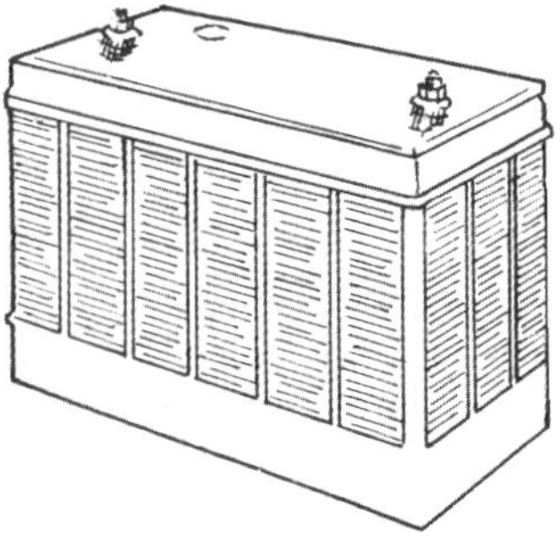

Figure 4-6. Battery for photovoltaic use. (*©ARCO Solar, Inc., a subsidiary of Atlantic Richfield Co.*)

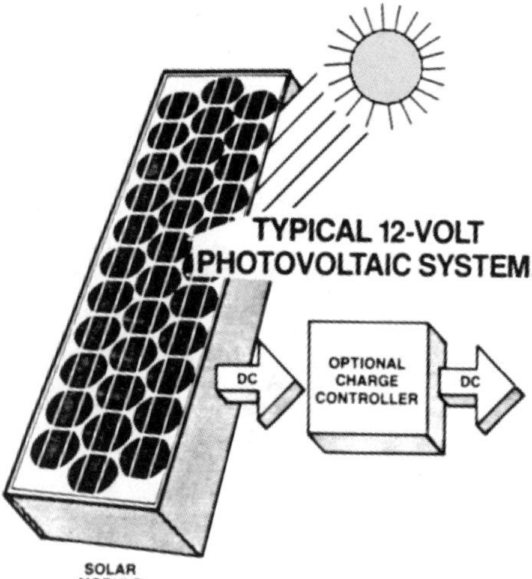

Figure 4-7. Typical photovoltaic system. (*©ARCO Solar, Inc., a subsidiary of Atlantic Richfield Co.*)

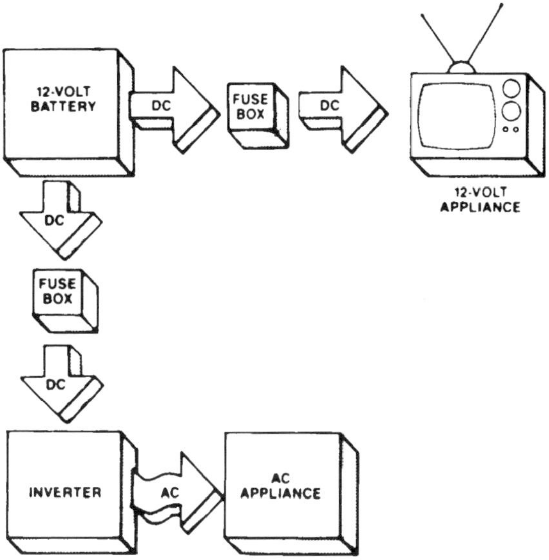

Figure 4-8. Typical photovoltaic system. (*©ARCO Solar, Inc., a subsidiary of Atlantic Richfield Co.*)

shown in Figures 4-7 and 4-8, which illustrate a typical 12-volt photovoltaic system. The production of electricity from direct sunlight allows power to be brought to remote areas for purposes ranging from emergency lighting to pumping water from a well in the desert. The applications of solar power are endless, and will continue to be used to provide power where other systems would be prohibitively expensive.

Photovoltaic Module

Figure 4-9 shows an ARCO Solar™ M81 Photovoltaic Module, which employs 33 electrically matched, single-cell crystal solar cells that directly convert sunlight into electricity. The module is suitable for a wide variety of low-power applications, such as powering navigation aids, telemetry relay stations, valve control, auxiliary equipment, recreational vehicles, and maintaining the charge on storage batteries.

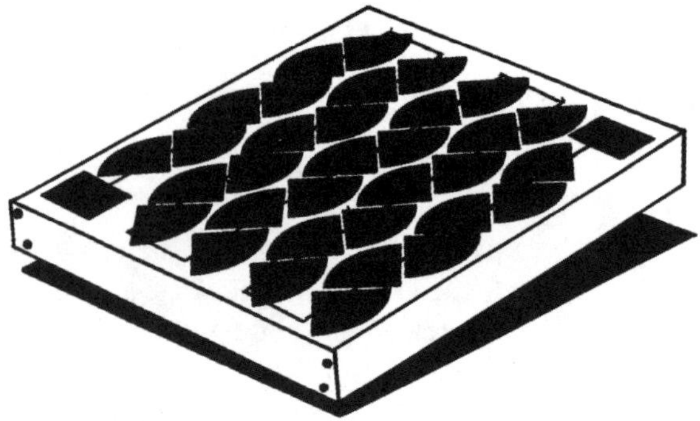

Figure 4-9. ARCO Solar™ photovoltaic module. (*©ARCO Solar, Inc., a subsidiary of Atlantic Richfield Co.*)

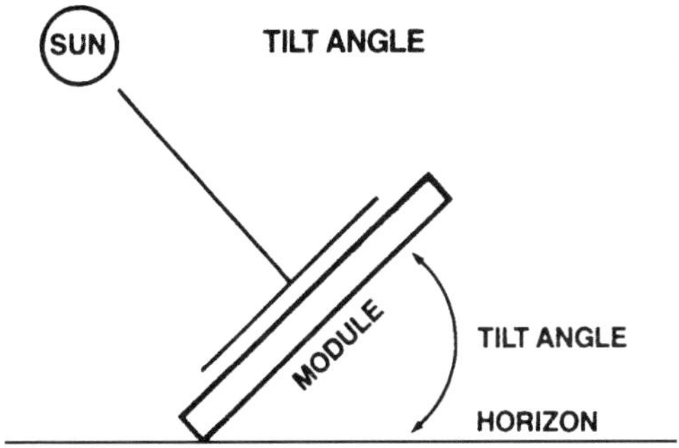

Figure 4-10. Tilt angle for solar module. (*©ARCO Solar, Inc., a subsidiary of Atlantic Richfield Co.*)

Tilt Angles

Figure 4-10 illustrates the tilt angle desired to ensure that maximum sunlight strikes the module(s). Consequently, the module must be tilted to be as perpendicular to the sun as possible at all times. Table 4-1 provides a suggested single position tilt angle for different locations throughout the world. (If your area is not listed, choose the closest city indicated on the chart.) Table 4-1 is presented here as an example only, and does not imply to be 100% accurate. It is based on a nominal yearly average of plus or minus 10% ampere-hours per day. The performance in any given day may be quite different.

(*text continued on page 113*)

Table 4-1
Performance Guide

*13.4V represents the charge voltage of a typical lead calcium battery operating @ 90% state-of-charge, 28°C temperature and a 2 amp charge rate.

Lat.	Location	Amp Hrs Generated By One M-61 Module Per Day	Tilt Angle
27N	Algeria, Aoulef	9.93	35S
30N	Algeria, Beni-Abbes	10.39	40S
34N	Algeria, Chottech Cherqui	9.24	55S
25N	Algeria, Quallen	10.40	25S
23N	Algeria, Tamanrasset	10.77	25S
30N	Arab Rep. Of Egypt, Giza	9.43	50S
41S	Argentina, Bariloche	6.44	65N
35S	Argentina, Buenos Aires	7.83	60N
28S	Argentina, Corrientes	8.13	50N
31S	Argentina, Rafaela	4.92	55N
35S	Australia, Adelaide	8.29	60N
35S	Australia, Canberra	8.63	60N
38S	Australia, Melbourne	7.02	60N
32S	Australia, Perth	8.67	55N

*©ARCO Solar, Inc., a subsidiary of Atlantic Richfield Co.

33S	Australia, Sydney	8.93	45N
19S	Australia, Townsville	9.67	25N
48N	Austria, Vienna	4.34	70S
17S	Bolivia, La Paz	10.36	25N
20S	Botswana, Maun	9.65	20N
59N	Canada, Churchill	6.81	80S
45N	Canada, Dartmouth	5.71	70S
54N	Canada, Edmonton	7.46	70S
62N	Canada, Fort Simpson	5.25	85S
46N	Canada, Montreal	5.73	70S
44N	Canada, Toronto	5.95	65S
49N	Canada, Vancouver	4.66	70S
12N	Chad, Ford Lamy	9.84	20S
23N	Chile, Atacama	12.77	30N
34S	Chile, Santiago	8.16	55N
44N	China, Changchun	7.40	60S
38N	China, Chefoo	7.65	55S
31N	China, Shanghai	6.76	55S
5N	Colombia, Bogota	7.38	15N
4S	Congo Republic, Brazzaville	6.92	25N
14N	El Salvador, San Salvador	9.83	20S
63N	Finland, Luonet Jarvi	3.98	85S
49N	France, Paris	4.87	70S
52N	Germany, Bochum	2.68	75S
48N	Germany, Munich	5.54	70S
6N	Ghana, Accra	7.50	15N
38N	Greece, Athens	7.11	60S
12N	Guinea-Bissau	9.35	20S
22N	Hong Kong	7.42	15S

(*table continued on next page*)

Table 4-1 (*Continued*)

Lat.	Location	Amp Hrs Generated By One M-61 Module Per Day	Tilt Angle
19N	India, Bombay (Mumbai)	9.54	15S
23N	India, Calcutta	8.59	15S
13N	India, Madras	9.66	20S
6S	Indonesia, Jakarta	6.82	15S
4S	Indonesia, West Irian	9.34	20N
32N	Iraq, Al-Kut	9.29	50S
32N	Israel, Jerusalem	10.20	50S
41N	Italy, Napoli	5.87	65S
46N	Italy, Palianza	6.28	70S
35N	Japan, Nagoya	7.73	25S
43N	Japan, Sapporo	5.39	65S
36N	Japan, Tokyo	5.17	40S
1S	Kenya, Nairobi	8.30	25N
35N	Korea, Puzan	9.20	25S
38N	Korea, Seoul	7.64	50S
34N	Lebanon, Ksara	9.05	55S
15N	Manana ls., Saipan	9.18	25S
17N	Mauritania, Nema	8.70	20S
20N	Mexico, Mexico City	9.23	15S
28N	Mexico, Nonoava	10.81	30S
23N	Mexico, Cabo San Lucas	9.04	35S
20N	Mexico, Tomatlan	9.10	25S
18N	Mexico, Tuxtla Guiterrex	7.20	15S
17N	Mexico, Acapulco	9.09	20S
28N	Morocco, Cabo Juby	8.81	35S

23S	Namibia (SWA), Windhoek	11.34	20N
30N	Nepal, Saga	8.56	15S
9S	New Guinea, Merauke	7.31	30N
4S	New Guinea, Rabul	8.11	20N
18S	New Zealand, Nandi	9.69	30N
29S	New Zealand, Raoul ls.	8.24	50N
41S	New Zealand, Wellington	6.48	65N
17N	Niger, Agadez	10.29	20S
7N	Nigeria, Benin City	5.94	15N
9N	Nigeria, Yola	8.14	15N
60N	Norway, Bergen	2.72	85S
25N	Pakistan, Karachi	8.69	15S
30N	Pakistan, Queta	10.52	45S
9N	Panama, Panama	7.16	15N
12S	Peru, Huancayo	11.68	20N
15N	Philippines, Quezon City	7.03	25S
38N	Portugal, Lisbon	8.77	60S
19N	Puerto Rico, San Juan	10.41	20S
26N	Saudi Arabia, Dhahran	9.74	35S
24N	Saudi Arabia, Riyadh	9.55	35S
9N	Sierra Leone, Longo	7.09	15N
1N	Singapore	6.56	15N
41N	Spain Madrid	7.33	65S
14N	Sudan, El-Fasher	9.20	20S
20N	Sudan, Port Sudan	9.37	25S
63N	Sweden, Froson	4.41	85S
56N	Sweden, Svalov	3.86	80S
25N	Taiwan, Taipei	5.35	50S

(*table continued on next page*)

Table 4-1 (*Continued*)

Lat.	Location	Amp Hrs Generated By One M-61 Module Per Day	Tilt Angle
23N	Taiwan, Tainan	8.53	15S
24N	Taiwan, Kwarenko	7.08	45S
14N	Thailand, Bangkok	8.17	20S
3N	Uganda, Moroto	10.31	15N
29S	Republic, So. Africa, Alexander Bay	11.26	45N
30S	Republic, So. Africa, Durban	8.41	35N
34S	Republic, So. Africa, Capetown	9.41	55N
24S	Republic, So. Africa, Pietersburg	10.52	25N
26S	Republic, So. Africa, Pretoria	10.13	25N
29S	Republic, So. Africa, Upington	10.42	25N
52N	United Kingdom, Cambridge	3.87	75S
United States of America			
32N	AL, Montgomery	7.55	55S
61N	AK, Bethel	6.41	60S
65N	AK, Fairbanks	5.19	90S
62N	AK, Matanuska	5.60	75S
33N	AZ, Phoenix	10.50	45S
35N	AR, Little Rock	7.29	60S
39N	CA, Davis	8.27	60S
37N	CA, Fresno	8.42	60S
34N	CA, Los Angeles	9.57	50S
34N	CA, Riverside	10.28	45S
40N	CO, Boulder	8.40	50S
40N	CO, Granby	10.27	40S

39N	D.C., Washington	6.74	60S
30N	FL, Gainsville	8.72	45S
26N	FL, Miami	9.25	30S
34N	GA, Atlanta	7.80	55S
21N	HI, Honolulu	9.74	35S
44N	ID, Boise	8.17	65S
42N	IL, Chicago	4.57	65S
40N	IN, Indianapolis	6.46	65S
42N	IA, Ames	7.10	65S
38N	KN, Dodge City	10.27	45S
39N	KN, Manhattan	7.35	60S
38N	KY, Lexington	8.09	60S
30N	LA, New Orleans	6.35	55S
32N	LA, Shreveport	7.32	55S
44N	ME, Portland	7.49	65S
42N	MA, Boston	5.92	65S
43N	MI, E. Lansing	6.29	65S
46N	MN, St. Cloud	7.50	70S
39N	MO, St. Louis	7.11	60S
47N	MT, Great Falls	7.85	70S
41N	NM, N. Omaha	8.09	60S
36N	NV, Las Vegas	10.89	45S
40N	NJ, Sea Brook	6.55	65S
35N	NM, Albuquerque	11.64	40S
41N	NY, New York City	6.26	65S
39N	NC, Ely	10.81	50S
36N	NC, Greensboro	7.43	60S
47N	ND, Bismark	8.51	65S

(table continued on next page)

Table 4-1 (*Continued*)

Lat.	Location	Amp Hrs Generated By One M-61 Module Per Day	Tilt Angle
41N	OH, Cleveland	6.08	65S
35N	OK, Oklahoma City	9.21	50S
42N	OR, Medford	7.20	65S
40N	PA, Pittsburgh	4.76	65S
41N	RI, Newport	6.68	65S
33N	SC, Charleston	7.95	55S
44N	SD, Rapid City	9.39	55S
36N	TN, Nashville	6.85	60S
32N	TX, Big Spring	9.20	45S
33N	TX, Fort Worth	8.95	50S
41N	UT, Salt Lake City	8.24	65S
38N	VA, Richmond	6.61	60S
47N	WA, Seattle	5.43	70S
48N	WA, Spokane	7.37	70S
38N	WV, Charleston	6.61	60S
43N	WI, Madison	6.88	65S
43N	WY, Lander	11.03	45S
46N	USSR, Odessa	5.68	70S
43N	USSR, Vladivostok	7.86	35S
60N	USSR, Leningrad	3.58	85S
35S	Uruguay, Montevideo	8.49	60N
10N	Venezuela, Barcelona	8.75	20S
11N	Venezuela, Maracaibo	8.67	20S
13N	Yeman, Aden	9.31	20S
45N	Yugoslavia, Beograd	5.64	70S

12S	Zaire, Lubumbash	8.20	15S
1N	Zaire, Kisangani	7.29	15N
20S	Zimbabwe, Bulawayo	10.16	15N

(*text continued from page 106*)

SUMMARY OF SOLAR MODULE SYSTEM

A complete photovoltaic system contains one or more solar modules, a structure to mount the modules, a battery storage system, and a charge control device to prevent the batteries from overcharging. In comparing solar modules, the most important consideration is the ability to charge batteries, not the number of cells. Increasing the number of cells might increase the voltage to the theoretical maximum power point, but past a certain point that voltage is not usable.

GLOSSARY OF TERMS USED IN SOLAR APPLICATIONS

AC (alternating current)—An electrical current that reverses from a positive to a negative voltage, changing at regular intervals and constantly changing in magnitude. Expressed in cycles per second or hertz, it is the current commonly used in household electricity.

Ampere or AMP—The standard unit for measuring the rate of flow of an electrical current. One ampere will flow when one volt is applied across a resistance of one ohm.

Array—More than one solar module joined together forming a group of modules.

Capacity—The amount of stored electricity in a battery, measured in amp hours.

Current—The net transfer of electric charge per unit of time propelled by voltage. Expressed in amperes.

Cycle life—A measure of how many times (cycles) a battery may be charged or discharged to a given state of charge until the capacity is reduced to 89% of original.

DC (direct current)—The direct or constant flow of electrical current in one direction, from the negative to the positive potential. Produced by batteries and by solar modules.

Load—The appliance consuming electricity generated by the solar system.

Module—The lowest denominator of encapsulated solar cells, another name for a single solar panel containing numerous cells.

Photovoltaic—Capable of generating a voltage as a result of exposure to light.

State of charge—A measure in percentage of how much of total capacity is contained in battery storage.

Solar cell—The basic unit of measuring electrical pressure that causes electricity to flow.

Watts—The unit of measuring electrical power, equal to the power developed in a circuit by current of one ampere flowing through a potential difference of one volt. (Simply, volts × amps = watts.)

BATTERIES FOR ELECTRONIC DEVICES

The performance of an electronic device is only as good as the battery inside it. The service life of high-performance batteries has been improved significantly in recent years, and it continues to improve. Today's batteries produce a high-energy output, are reliable, have a long shelf life, and are excellent in low-temperature applications. Detailed technical information on all types of batteries is readily available from the various battery manufacturers.

Types of Batteries

Alkaline-manganese dioxide, mercury oxide, silver oxide, zinc air, and lithium/manganese dioxide batteries are compared in Table 4-2, and shown in Figure 4-11. (Refer to the Glossary of Terms Used in Battery Applications.)

Storing Batteries

Extra batteries should be stored in a dry place at normal room temperature. Refrigerating batteries is not necessary. Extreme heat or cold reduces battery performance. If batteries must be stored in either hot or cold conditions, alkaline batteries will generally perform better in these conditions than regular carbon batteries.

(*text continued on page 118*)

Table 4-2
Battery Characteristic Comparison

Principal Chemical Systems	General Characteristics (under optimum operating conditions)	Configurations and Rated Capacity*	Nominal Cell Voltage
Alkaline-Manganese Dioxide (Zn/MnO_2)	Popular, multi-use premium battery; good low temperature and high rate performance; good shelf life; sloping discharge characteristic.	Cylindrical cells to 20 Ah; also available in small button cells.	1.5
Mercuric Oxide (Zn/HgO)	High volumetric capacity; good shelf life; flat discharge characteristic.	Button and cylindrical cells from 40 mAh to 13 Ah; also available in special larger sizes.	1.35
Silver Oxide (Zn/Ag_2O)	High gravimetric capacity; good shelf life; flat discharge characteristic.	Button cells to 180 mAh for specialized applications.	1.5
Zinc Air (Zn/O_2)	Highest energy density on continuous discharge; excellent shelf life (unactivated); moderate rate capability limited shelf life when activated; flat discharge characteristic.	Button cells to 6.6 Ah.	1.4
Lithium/ Manganese Dioxide (Li/MnO_2)	High energy density; excellent rate capability and low temperature performance; excellent shelf life; relatively flat discharge characteristic.	Coin cells to 550 mAh; spiral-wound cylindrical cells to 1.3 Ah; cylindrical (bobbin-type) cells to 1.9 Ah.	3.0

*Courtesy of Duracell, Inc.

Principal Chemical Systems		Energy Density				Typical Applications	Shelf Life (% loss/year @ 70°F)	Typical Operating Temperature
		Weight		Volume				
		Wh/lb.	Wh/kg	Wh/in.^{3}	Wh/l			
Alkaline-Manganese (Zn/MnO_2)	Button:	25	55	2.4	145	General purpose, e.g., lighting, photographic equipment, toys, radio and T.V., tape recorders, watches and instruments. Popular for high drain applications.	3%	−4°F to 130°F (−20°C to 54°C)
	Cylindrical:	60	130	5.8	350			
Mercuric Oxide (Zn/HgO)	Button:	50	110	7.3	445	Hearing aids, medical and photo-devices, pagers, detection and emergency equipment, and other applications requiring steady output voltage.	4%	15°F to 130°F (−9°C to 54°C)
	Cylindrical:	55	120	7.4	450			
Silver Oxide (Zn/Ag_2O)	Button:	60	130	8.2	500	Hearing aids, watches, photographic equipment, and special electronics requiring small, high capacity batteries.	6%	−4°F to 130°F (−20°C to 54°C)
Zinc Air (Zn/O_2)	Button:	140	310	19.0	1,150	Hearing aids, medical monitoring instruments, pagers, hand-held portable two-way radios, telephones, emergency lighting, and other frequent use applications.	2% (unactivated)	32°F to 122°F (0°C to 50°C)
Lithium/Manganese Dioxide (Li/MnO_2)	Coin:	80	175	8.3	505	General purpose in photographic and electronic applications requiring small, high power batteries, e.g., calculators, lighting, clock/calendar, memory-backup, photoflash, motor-driven cameras, medical equipment and instruments.	0.5–1%	−40°F to 140°F (−40°C to 60°C)
	Cylindrical:							
	Spiral-Wound:	105	230	8.3	505			
	Bobbin:	135	300	10.1	615			

Courtesy of Duracell, Inc.

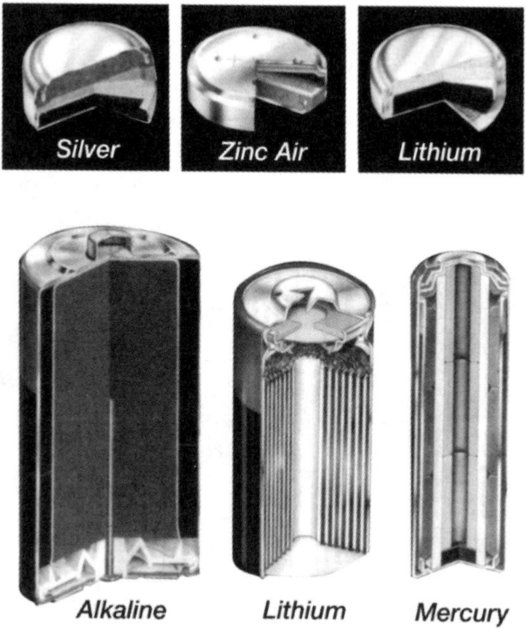

Figure 4-11. Primary battery systems. (*Courtesy of Duracell, Inc.*)

(*text continued from page 115*)

Safety Tips

- Keep battery contact surfaces clean by wiping them off with a pencil eraser or rough cloth.
- Never dispose of batteries in a fire, as the batteries may explode.

- Don't carry loose batteries in your pocket or purse, as the batteries can be shorted by contact with metal objects, and may leak or rupture.
- Never attempt to recharge a battery unless it is specifically marked "rechargeable."
- Replace batteries in your equipment with the manufacturer's suggested type and size.
- Always make sure that the positive and negative terminals on the battery are aligned properly when inserting them into your equipment.
- Remove batteries from equipment that will not be used for several months or longer.

GLOSSARY OF TERMS USED FOR BATTERIES*

Ambient Humidity

The average humidity of the surroundings.

Ambient Temperature

The average temperature of the surroundings.

Ampere-Hour Capacity

The quantity of electricity measured in ampere-hours (Ah) which may be delivered by a cell or battery under specified conditions.

Anode

The electrode in an electrochemical cell where oxidation takes place. During discharge, the negative electrode of the cell is the anode.

*Courtesy of Duracell, Inc.

Battery

Two or more electrochemical cells electrically interconnected in an appropriate series/parallel arrangement to provide the required operating voltage and current levels. Under common usage, the term "battery" is often also applied to a single cell.

Bobbin

A cylindrical cell design utilizing an internal cylindrical electrode, and an external electrode arranged as a sleeve inside the cell container.

C Rate (also see Hourly Rate)

Discharge or charge current, in amperes, expressed in multiples of the rated capacity. For example, C/20 discharge current for a battery rated at 5-h discharge rate is:

$$\frac{C_5}{20}\,(\text{AH}) = \text{current (A)}$$

As a cell's capacity is not the same at all discharge rates and usually increases with decreasing rate, a discharge at the $C_5/20$ rate will run longer than 20 hours.

Capacity

The total number of ampere-hours or watt-hours that can be withdrawn from a fully charged battery under specified conditions of discharge.

Capacity Retention

The fraction of the full capacity available from a battery under specified conditions of discharge after it has been stored for a period of time.

Cathode

The electrode in an electrochemical cell where reduction takes place. During discharge, the positive electrode of the cell is the cathode.

Cell

The basic electrochemical unit used to generate or store electrical energy.

Closed-Circuit Voltage (CCV)

The potential or voltage of a battery when it is discharging.

Continuous Test

A test in which a battery is discharged to a prescribed end-point voltage without interruption.

Coulomb

The amount of electricity transported by a current of one ampere flowing for one second.

Current Collector

An inert member of high electrical conductivity used to conduct current from or to an electrode during discharge or charge.

Current Density

The current per unit active area of the surface of an electrode.

Cutoff Voltage

The battery voltage at which the discharge is terminated. The cutoff voltage is specified by the battery manufacturer and is generally a function of discharge rate.

Depth of Discharge

The ratio of the quantity of electricity (usually in ampere-hours) removed from a battery to its rated capacity.

Discharge

The conversion of the chemical energy of a battery into electrical energy, and the withdraw of the electrical energy, usually in order to operate and electronic device.

Discharge Rate

The rate, usually expressed in amperes, at which electrical current is taken from the battery.

Drain

The current withdrawn from a battery during discharge.

Dry Cell

A cell with immobilized electrolyte. The term "dry cell" is often used to describe the Leclanche cell.

Duty Cycle

The operating regime of a battery including factors such as charge and discharge rates, depth of discharge, cycle length, and length of time in the standby mode.

Electrochemical Equivalent

Weight of a substance that is deposited at an electrode when the quantity of electricity which is passed is one coulomb.

Electrode

The site, area, or location at which electrochemical processes take place.

Electrolyte

The medium which provides the ion transport mechanism between the positive and negative electrodes of a cell.

End Voltage

The prescribed voltage at which the discharge (or charge, if end-of-charge voltage) of a battery may be considered complete (also cutoff voltage).

Energy

The output capability of a cell, usually expressed in watt-hours, which is the ampere-hour capacity multiplied by the average discharge voltage.

Energy Density

The ratio of the energy available from a battery to its volume (Wh/L) or weight (Wh/kg).

Forced Discharge

Discharging a battery, with external battery or power source, below zero volts into voltage reversal.

Gassing

The evolution of gas from one or more of the electrodes in a cell. Gassing commonly results from local action (self-discharge) or from the electrolysis of water in the electrolyte during charging.

Hertz

The standard unit of frequency. A frequency of one complete cycle per second is a frequency of one hertz.

Hourly Rate (also see C RATE)

A discharge rate, in amperes, of a battery which will deliver the specified hours of service to a given end voltage.

Intermittent Test

A test during which a battery is subjected to alternate periods of discharge and rest according to a specified discharge regime.

Internal Impedance

The opposition exhibited by a circuit element (cell or battery) to the flow of an alternating current (a.c.) of a particular frequency as a result of resistance, induction and capacitance.

Internal Resistance

The opposition exhibited by a circuit element to the flow of direct current (d.c.). In a cell, the internal resistance is the sum of the ionic and electronic resistances of the cell components.

Limiting Current

The maximum current drain under which the particular battery will perform adequately under a continuous drain. The rate is based on whatever drain rate reduces the running voltage to 1.1 volts.

Midpoint Voltage

The voltage of a battery midway in the discharge between the fully charged state and the end voltage.

Nominal Voltage

The characteristic operating voltage or rated voltage of a battery (as distinct from Working Voltage, Midpoint Voltage, etc.)

Open-Circuit Voltage (OCV)

The difference in potential between the terminals of a cell or voltage when the circuit is open (no-load condition).

Parallel

Term used to describe the interconnection of cells or batteries in which all the like terminals are connected together.

Positive Temperature Coefficient (PTC) Device

A thermally reactive device which becomes highly resistive at a specific temperature.

Polarization

The lowering of the potential of a cell or electrode from its equilibrium value caused by the passage of an electric current.

Primary Battery

A battery which is not intended to be recharged and is discarded when the battery has delivered all its electrical energy.

Pulse Current

A periodic current drain of higher than normal drain rates.

Rated Capacity

The number of ampere-hours a battery can deliver under specific conditions (rate of discharge, end voltage, temperature); usually the manufacturer's rating.

Reversal

The changing of the normal polarity of a battery due to overdischarge.

Safety Vent

A venting mechanism designed into a cell which activates under specific conditions of abuse to relieve internal pressure.

Self-Discharge

The loss of useful capacity of a battery in storage due to internal chemical action (local action).

Separator

An ionic permeable electronically nonconductive spacer which prevents electronic contact between electrodes of opposite polarity in the same cell.

Series

The interconnection of cells in such a manner that the positive terminal of the first is connected to the negative terminal of the second, and so on.

Service Life

The period of useful life of a primary battery before a predetermined end-point voltage is reached.

Shelf Life

The duration of storage under specified conditions at the end of which a battery still retains the ability to give a specified performance.

Short-Circuit Current (SCC)

The initial value of the current obtained from a battery in a circuit of negligible resistance.

Spiral-Wound

An electrode structure of high surface area created by winding the electrodes and separator into a spiral-wound, jelly-roll configuration.

Voltage Delay

Time delay for a battery to deliver the required operating voltage after it is placed under load.

Working Voltage

The typical voltage or range of voltage of a battery during discharge (also called operating voltage or running voltage).

5
Miscellaneous Electrical and Instrumentation Items

CABLE AND WIRE TIES

Wires and cables go through a maze of paths in a project—up and down, around and inside of tight corners, and then the same over again. Figure 5-1 shows a few of the types of cable ties used to anchor the wires and cables to each other and to objects in the path.

WIRE AND TERMINAL MARKERS

Thousands of self-adhesive strips printed with numbers, letters, symbols, etc. are mounted on dispenser cards. There are many similar type markers for voltage ratings, tape for color coding, safety stripes, and aisle marking. Figure 5-2 depicts some of the types of markers and their use. Wider

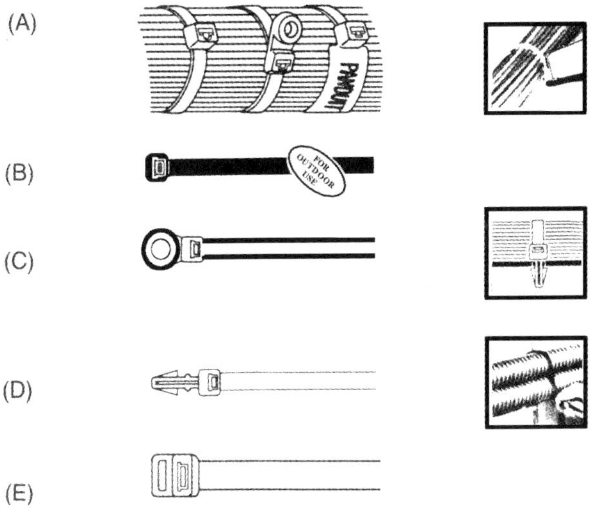

Figure 5-1. Cable ties: Panduit Pan-Ty® nylon ties (A); black weather-resistant cable ties (B); PCL clamps (C); lashing ties (D); and push mount ties (E). (*Courtesy of Jaqua-McKee, Inc.*)

strips and different methods are available for marking wires, cable, conduit, telephone line, systems, etc.

WIRE CONNECTORS

Figure 5-3 illustrates only a few of the many styles of wire connectors. Crimp type connectors are used for small

(text continued on page 132)

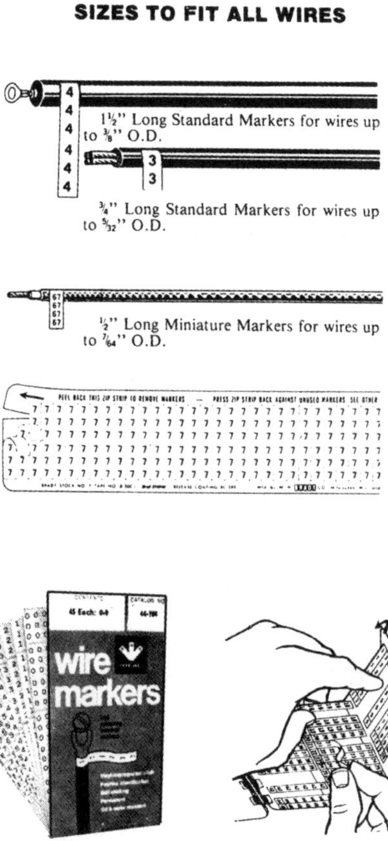

Figure 5-2. Wire and terminal markers. (*Courtesy of Jaqua-McKee, Inc.*)

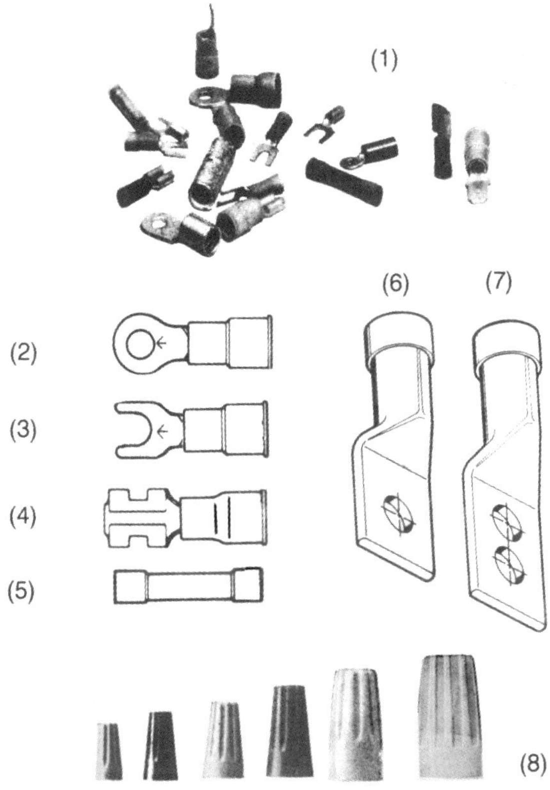

Figure 5-3. Wire connectors: crimp type (1); Scotchlok® ring tongue terminal (2); spade tongue (3); female quick-slide connector (4); butt connector (5); Scotchlok aluminum one-hole lug (6); two-hole lug (7); Ideal wire-nut connectors (8). (*Courtesy of Jaqua-McKee, Inc.*)

(text continued from page 129)

wire gauges. Several styles of the connectors are shown in (1) and enlarged in the other views. The connectors are color coded for size range, and have internal barrel serrations to grasp the wire strands firmly. A crimping tool is used to apply the connectors. In Figure 5-3, (2) is a Scotchlok ring tongue terminal; (3) is a spade tongue; (4) is a female quick-slide connector; (5) is a butt connector; (6) is a Scotchlok® aluminum one-hole lug; (7) is a two-hole lug; and (8) shows Ideal wire-nut connectors that are simply twisted on the ends of two wires to be joined, and provide a tight wire grip. The nuts are identified for size by color coding.

SPLICING KITS

Larger cables require more elaborate splicing methods. Figure 5-4A is the 3M 5400 series quick-splice incline splicing kit. The kit contains everything needed to make a splice that consists of a slip-on insulation body, and two end caps. The splice insulates and seals for a water proof connection for direct burial. Figure 5-4B is the Scotchcast 82-A1, 2, 3 splicing kit used for splicing nonshielded cables rated up to 5KV. (Multi-conductor cables rated up to 600 volts.) The kits have a two-piece, transparent snap-together mold body. The ends are sealed, and then a resin is poured into the mold until it rises in the funnel. After 30 minutes, the splice is ready. Figure 5-4C shows Scotch brand 5701, 1, 3 K Tape termination kits. These kits con-

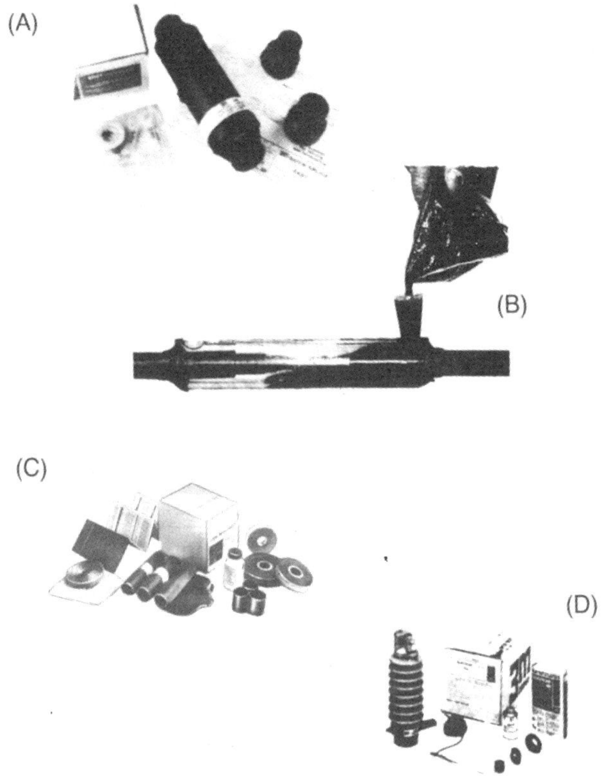

Figure 5-4. Wire splicing kits. (*Courtesy of Jaqua-McKee, Inc.*)

tain enough tape to make three terminations, and are designed to terminate cables from 5 to 15 KV. Figure 5-4D is the 3M porcelain termination kit. The kits are designed to terminate single-conductor cables rated from 5 to 25 KV. They contain a wet process porcelain insulator and materials for the termination.

THERMOCOUPLES AND ACCESSORIES

The measurement of temperature is one of the most vital functions in the control of manufacturing and processing operations. At the heart of all temperature measurements is the thermocouple. The concept of all thermocouples is the same. Two wires of dissimilar metal are joined together at one end. An increase in temperature creates an electronic force (emf) or voltage, which is transmitted through these wires to a monitoring device that reads this signal and displays it on a previously calibrated meter or digital device. Figure 5-5 shows the thermocouple and some of the accessories used in a system:

A. The thermocouple element with a t.c. head, and thermocouple wire (extension cable).
B. A thermowell. (The thermocouple element screws directly into the thermowell either in a process line or process vessel.)

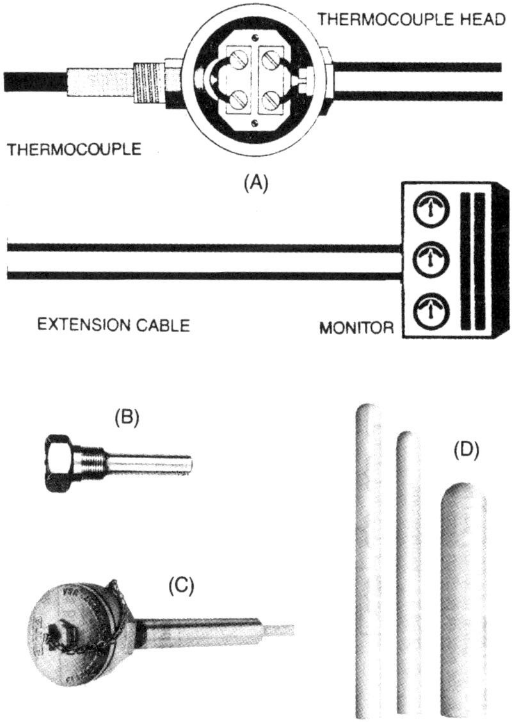

Figure 5-5. Thermocouple and thermowells. (*Courtesy of Alpha Wire Corp.*)

C. Thermocouple head on pipe nipple. (The element is inside the pipe nipple. Base metal thermocouple.)
D. Ceramic protection tubes for use in very high temperature areas. Figure 5-6 shows insulators for thermocouple wire used in high temperature areas.

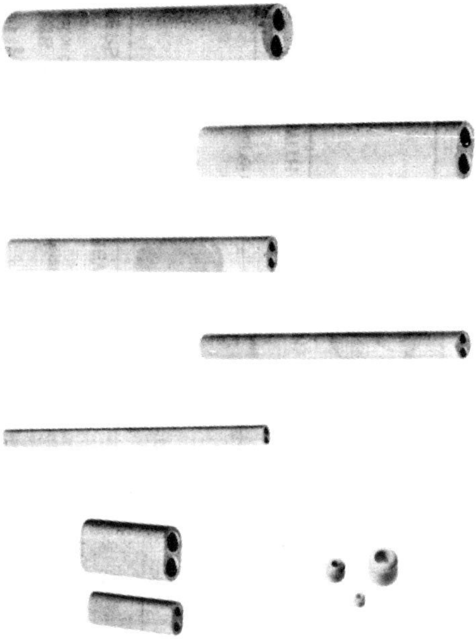

Figure 5-6. Thermocouple insulators for elevated temperatures. (*Courtesy of Alpha Wire Corp.*)

E. An alternative to ceramic thermocouple connectors are molded-on Zenite™ 6130 thermocouple connectors from DuPont. These are color coded to match the thermocouple in the circuit. They sell for $7 to $9 less than ceramic thermocouple connectors. Figure 5-7 shows thermocouples made from liquid crystal polymer resin.

Figure 5-8 shows more thermocouple items plus connectors and connector blocks. Make sure all small items for thermocouples are tagged with their part number (leave in manufacturer's carton if possible), and your commodity code number. Thermocouple assemblies are made from

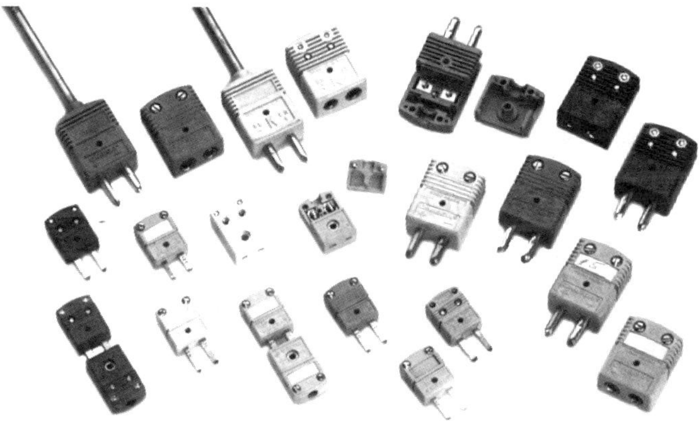

Figure 5-7. Low-cost, high-temperature Zenite™ LCP thermocouples. *(©Copyright Omega Engineering, Inc. All rights reserved. Reproduced with permission of Omega Engineering, Inc., Stamford, CT 06907.)*

Figure 5-8. Thermocouple sensors and accessories. (*Courtesy of Alpha Wire Corp.*)

brass, carbon steel, stainless steel, and monel. Thermocouples will be engineered and tagged for a specified project location.

GAUGE GLASSES

Level gauges mount on the outside of vessels usually on a bridle with other instrument items attached as well by pipe nipples and valves. The three liquid level gauges with transparent windows are types TL, TCL, and TM (Figure 5-9). These gauges will be tagged for a specific unit or application.

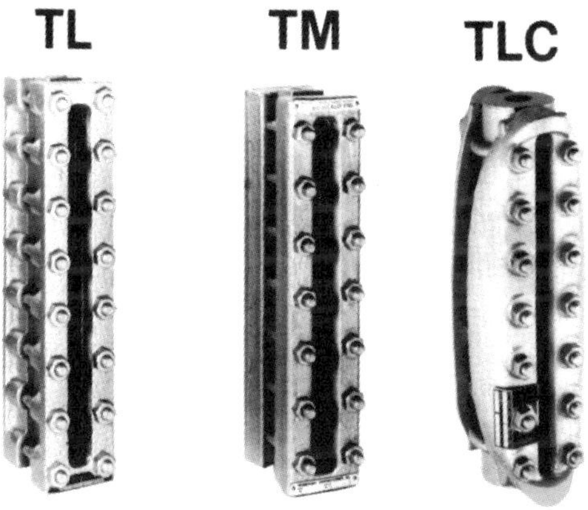

Figure 5-9. Gauge glasses. (*Courtesy of Richardson Engineering Services, Inc.*)

VIBRATION CONTROL

Fittings such as those shown in Figure 5-10 are used to control vibration on pumps, motors, and other applications in the process system. Illustrated are:

1. Stainless steel hose and braiding. Carbon steel nipples and braid ferrules are standard in this type of flexible hose connector.
2. A stainless steel braided connector with plate carbon steel flanges.
3. The expansion joint has heavy pipe attached to a stainless steel bellows.
4. The same expansion joint described in (3) but this one has carbon steel plate flanges.

THERMOMETERS

Only a few thermometers are shown in Figure 5-11. The types shown are most common and typical for use on vessels. Thermometers are described by their range (i.e., 30 to 240°), size of dial (i.e., 4-in.), and style of connection (i.e., back connection, adjustable angle, bottom connection, and flexible extension). Note in Figure 5-11 the element is screwed into the thermowell on the thermometer on the right.

(text continued on page 143)

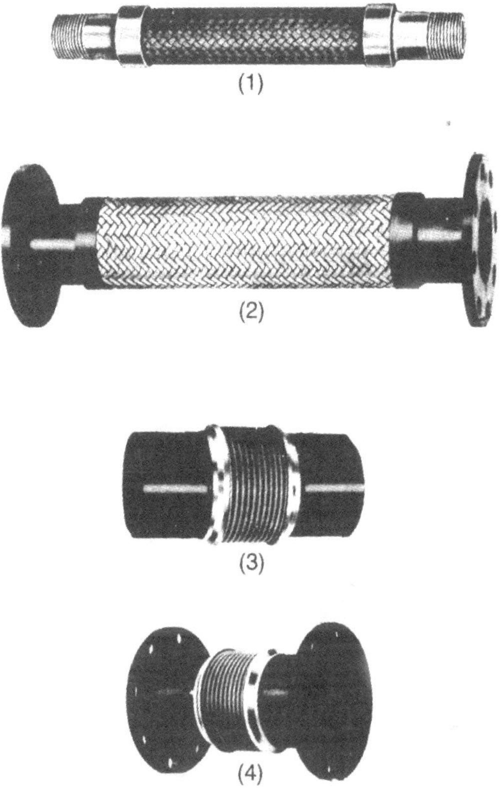

Figure 5-10. Flexible pump connectors and expansion joints. (*Courtesy of Richardson Engineering Services, Inc.*)

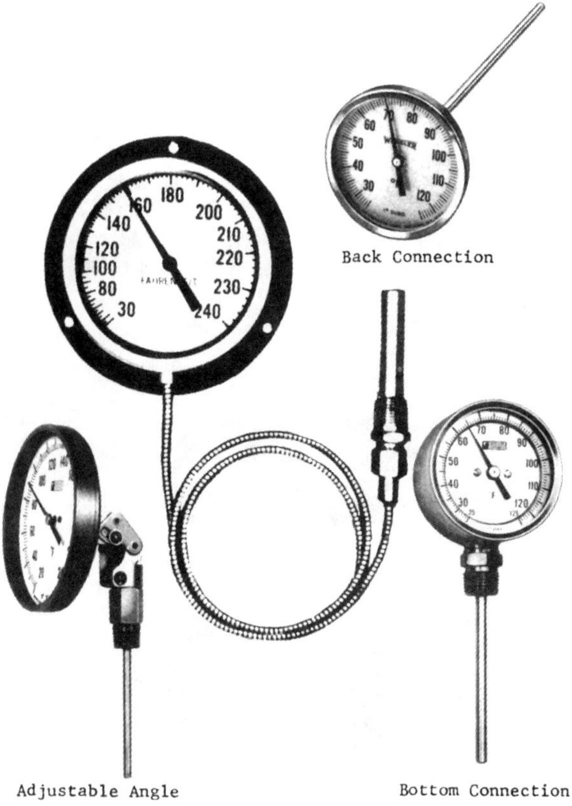

Figure 5-11. Thermometers. (*Courtesy of Richardson Engineering Services, Inc.*)

PRESSURE GAUGES

Figure 5-12 shows only five of the hundreds of styles and types of pressure gauges. Gauges are described by their pressure range (i.e., 0 to 15,000 psi), the size of the dial (i.e., 4½ in.), the type of case (i.e., brass or black steel), and the bourdon tube, which is the heart of a gauge, being a

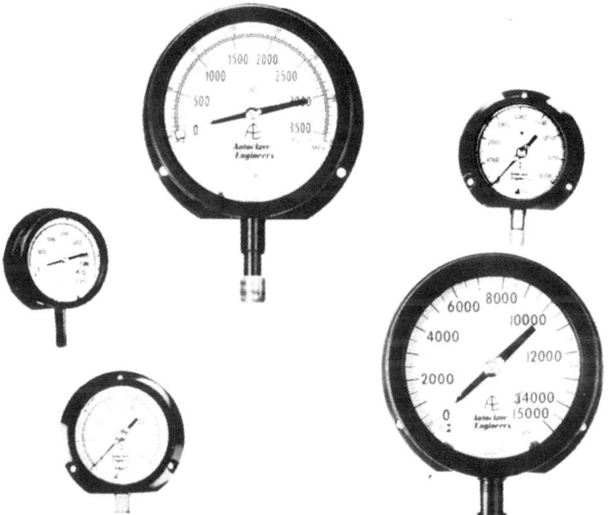

Figure 5-12. Pressure gauges. (*Courtesy of Autoclave Engineers, Inc.*)

curved tube with a flattened cross section and closed on one end. The crystal may be made of acrylic, glass, laminated safety glass, or other types of clear product. The connection size and the mounting options can be bottom mount, back mount, etc. Some gauges are equipped with a blow out back for pressure relief in case of bourdon tube failure. Some have a pointer zero adjustment located on the front of the gauge behind the dial cover. Gauges can be equipped with adjustable low and high electrical contacts to provide pressure control for automatic or remote operation, or for fail-safe points.

Store the gauges in their manufacturer's cartons. Never apply oil or other chemicals to the threads or any other parts of the gauge. The gauges will be engineered for a specific location in the project, and will be tagged as such. Only the extra gauges for spares will not be tagged to specific locations. Check the dial face for damage when gauges are received. Never issue a gauge for use in pressures that may exceed the ratings.

EXPENDABLE ITEMS

One of the very first material shipments to arrive at a project when process unit construction has commenced are the expendable electrical items. The materials, after the receiving formalities have been accomplished, should then be submitted to the craft foreman for his signature and reissued to the electricians. (On large projects, this system may not be feasible. Request advice from the project manager through your supervisor.)

Expendables may consist of various quantities of the following materials:

Hex head cap screws, lockwashers, flatwashers, hex nuts
Wire terminals, tongue and blade
Electrical tape
Glyptol and thinner
Wire markers
Thread cutting oil
Insulating tape
Scotchlok®, types Y and R
Chico A3 and Chico Fibre
Duxseal
Wire pulling compound
Ty-rap
Cans thread lubricant

Bills of materials may vary from project to project for expendable materials. They are a very important part of any project.

6
Bar Coding Systems

Bar coding systems are widely used throughout the world for accuracy, cost reduction, speed, accounting, time keeping, material control, identification, and tracking, and are very flexible for adding your own applications.

For materials control in field warehouses, the bar codes may be prepared along with purchase orders from data stored in a text file and retrieved to print whenever required.

Numbering systems or commodity code numbers already existing in the users' archives may already be used in your facility, but if not, you may consider the uniform-universal system for the construction industry worldwide that is suggested later in this chapter. The flexibility in preparing your bar code labeling allows you to have full control of the size, design, and layout of the labels or plates. In addition to the bar code itself, labels can include text, graphics, radio frequencies, and fields merged from data files. You'll find more details about bar coding on the Internet. There are other information sources such as books, trade magazines, trade shows, and even your favorite computer store.

"Bumpy bar codes" have been created by Norcom Electronics Corporation. These codes are ideal for harsh environments, and are virtually indestructible during sandblasting and painting. They stand up to high temperatures, moisture, solvents, and chemical baths, and can last the life of the object as permanent identification. The 3-D bumpy bar codes are scanned by a bar code reader that "reads" the height differences between the bars and spaces rather than traditional character contrast. Norcom's Endure 1800 Series machine can emboss 15,000 characters per hour onto metal plates of various sizes. Examples of these plates are shown in Figure 6-1.

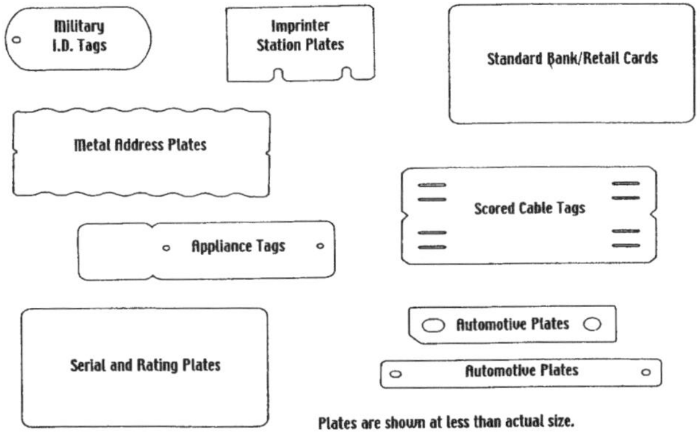

Figure 6-1. Variety of plate sizes. (*Courtesy Norcom Electronics Corp.*)

To run a bar codes system on your PC, the latest Windows software will probably suffice. However, it will be wise to consult your dealer about selecting a bar code system for your applications. Software is available to print label bar codes on any laser jet printer for normal use. Bar codes may be printed on any suitable printer for testing and practice.

A simple system for entry level tracing, inventory, shipping/receiving, and document control is used as an introduction to bar coding. TimeWand II® offers suitable memory, programmability, and hardware features for demanding applications. This unit is is available in 128K and 320K. The unit is the size of a credit card and weighs 2 ounces (see Figure 6-2). A fully portable bar code reader, the Videx® TimeWand® can be used as a remote system, as shown in Figure 6-3.

Figure 6-2. TimeWand® portable bar code reader. (*Courtesy of Videx®, Inc.*)

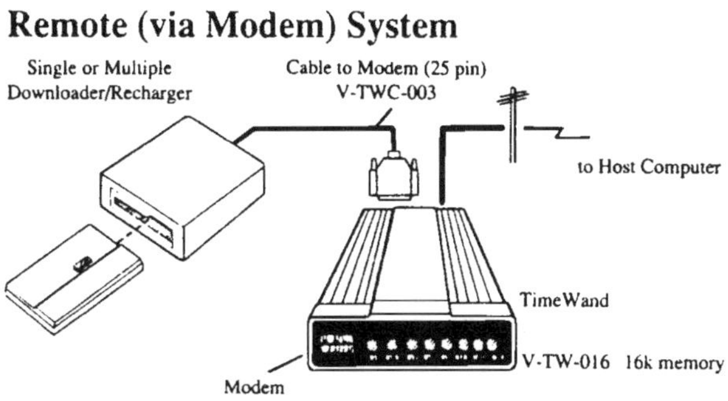

Figure 6-3. Remote bar code system via modem. (*Courtesy of TimeKeeping Systems, Inc.*)

These small bar code readers will fit in a shirt pocket. Each contain an internal clock, and time-stamp each code as it is read. Because the communication is keyless, the scanned information is error free (see Figure 6-4). On a project, different craftsmen could use these types of code readers while performing their varied duties.

As bar codes are read with the TimeWand®, a beeper indicates a good or bad read, and whether the TimeWand® is nearly full or full of data. It can operate up to three days without recharging, and can be recharged overnight in the Downloader/Recharger units, as shown in Figure 6-5.

Code data stored in the TimeWand® is transferred to a PC by means of the Downleader/Recharger unit. For use by various employees in a large fabricating facility, a multiple

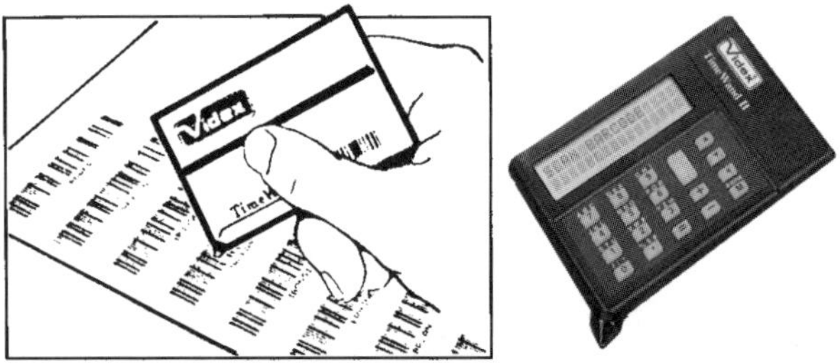

Figure 6-4. Portable TimeWand II® bar code reader. (*Courtesy of Videx®, Inc.*)

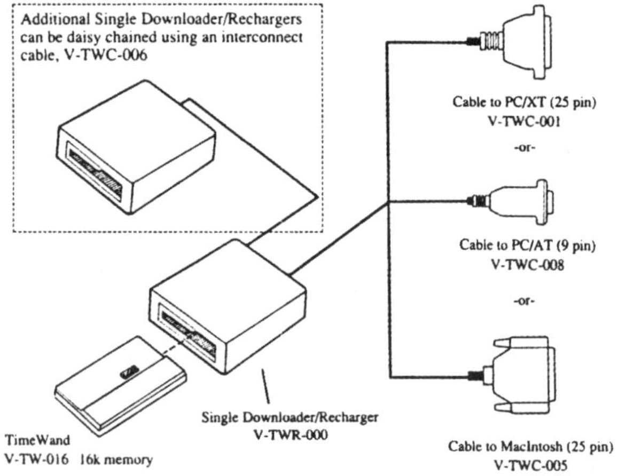

Figure 6-5. Single TimeWand® System. (*Courtesy of TimeKeeping Systems, Inc.*)

Downloader/Recharge unit, consisting of a controller and up to 25 add-ons, can be downloaded via one serial port (see Figure 6-6). After downloading, the TimeWand® can be stored overnight for recharging, and the various reports printed from the PC as required.

Bar coding is exciting and has the potential for many different applications in the future, such as for picking and placing parts, keeping medical records, tracking waste dis-

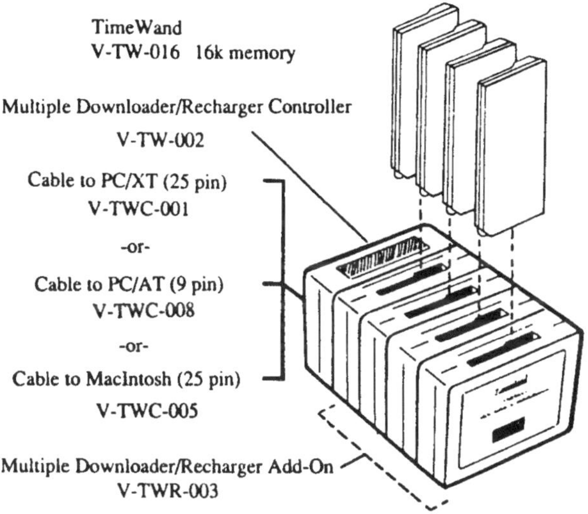

Figure 6-6. Multiple TimeWand® System. (*Courtesy of TimeKeeping Systems, Inc.*)

posal, evaluating currency exchange, recording debit cards, and on and on, even to invisible asset marking for moving and tracking property within your complex, or for tracing stolen assets.

Numbering systems can vary to include a numeric and verbal description on the bar code, to include either description alone, or to include a combination of the two. An industry-wide universal numbering system would contribute to uniformity in the industry. The numbering system could use steel specifications, tag numbers of engineered items, and other miscellaneous numbers.

SUGGESTED BAR CODE NUMBERING SYSTEM FOR CONSTRUCTION TRADES

To create uniformity, one suggested bar code numbering system would be to use the first digit or letter of the bar code to identify some type of steel, an item, or a combination of both. The next one or two digits would identify what the item is, such as a valve or a flange. Additional digits or letters would then show the schedule or thickness, bore size, or pressure rating, as required, and the last digits would indicate the size of the item, such as "12" for something that was 12 in. in diameter. The various systems can also be programmed to show metric equivalents. Bar codes may be made with as many numbers and letters as the user requires, but ten or less is suggested.

As an example, a typical bar code for a 6-inch gate valve, 150-lb RF, having an assigned V-number for a certain project, would be as follows: From Table 6-1, the first digit for carbon steel is "0." The second digit is the code for a valve, again "0." The next three numbers and letters show the valve's V-number, in this case V58. The last two digits indicate the size of the valve, which in this example is 12 in. Our number, therefore, would be 00V5812 and, if desired, a verbal description could also be embedded in the bar code.

Table 6-1
Number Codes for Bar Codes

Symbol	Material
0	Carbon Steels
1	Special Carbon Steels
2	Alloy Steels (Exotics)
3	304 Stainless Steels
3L	304L Stainless Steels
4	Unassigned
5	Unassigned
6	316 Stainless Steels
6L	316L Stainless Steels
7	Unassigned
8	Miscellaneous
PL	Structural Plates
SH	Structural Shapes

(table continued on next page)

Table 6-1 (*Continued*)

Secondary Identification Symbols

Symbol	Material
SP	Structural Pipe
ST	Miscellaneous Structural
SX	Expanded Metals
0	Valves
1	Pipes
2	Flanges
3	Fittings
4	Bolts
5	Socketweld Fittings
6	Screwed Fittings
7	Gaskets
8	Miscellaneous

Additional Identification Symbols

Symbol	Type
80	Schedule 80
125	125 psig
150	150 lb.
160	Schedule 160
1500RTJ	1,500-lb. Ring-type Joint
.500	.500-inch wall thickness
FP	Full Port
PI239	Pressure Indicator #239
RF	Raised Face
WF	Wide Flange

Suggested Size Codes

Symbol	Size
21	¼″
31	⅜″
41	½″
51	⅝″
61	¾″
01	1″
14	1¼″
15	1½″
02	2″
25	2½″
03	3″
	and continuing

Appendix

Abbreviations for Instrumentation Items

AC Air to close
AO Air to open
AT Analyzer transmitter
AY Analyzer computer
AE Analyzer element
BVD Blowdown valve
CC Corrosion coupon
ESD Emergency shutdown
FC Flow controller
FCV Flow control valve
FI Flow indicator
FIC Flow indicating controller
FR Flow recorder
FRC Flow recording controller
FT Flow transmitter
FE Flow element
FAH Flow alarm, high
FSH Flow safety, high
FG Sight glass
FIT Flow indicating transmitter
FAL Flow alarm low
FQY Totalized flow computer
FS Flow switch
FY Flow relay
FV Flow valve
FM Flow meter
FQI Flow totalizing indicator
HS Head switch
HV Hand valve
HCV Hand control valve
LAL Level alarm, low
LAH Level alarm, high
LC Level controller
LCV Level control valve

LG Level gauge
LI Level indicator
LIC Level indicating controller
LS Level switch
LT Level transmitter
PDM Positive displacement meter
PAL Pressure alarm, low
PAH Pressure alarm, high
PC Pressure controller
PCV Pressure control valve
PDI Differential pressure indicator
PDR Differential pressure recorder
PI Pressure indicator
PR Pressure recorder
PS Pressure switch
PSE Pressure safety element (rupture disc)
PSH Pressure safety, high
PSL Pressure safety, low
PSV Pressure safety valve
PT Pressure transmitter
PY Pressure relay
QX Event indicator
RO Restriction orifice
ROV Remote-operated valve
SAH High-speed alarm
SC Speed controller
SD Shutdown
SDV Shutdown valve
SI Speed indicator
SP Specialty item
SV Solenoid valve
SDY Shutdown relay
TAH Temperature alarm, high
TAL Temperature alarm, low
TC Temperature controller
TE Temperature element
TI Temperature indicator
TIC Temperature indicating controller
TR Temperature recorder
TT Temperature transmitter
TW Thermowell
VSH Vibration safety, high (low)
WR Wind recorder
XI Wind direction indicator
XT Wind direction transmitter
ZI Position indicator
ZS Position switch

Index

About the Author

R. R. Lee is the vice-president/owner of Lee's Material Services, Inc. in Houston, Texas, which received the prestigious Presidential "E" Award for excellence in export. Mr. Lee's 35 years' experience includes serving as material takeoff section leader with McDermott Engineering, purchasing agent, material/inventory control supervisor, and specialist/manager with Brown & Root International, Texaco, and Oasis Oil Company of Libya.